Designing **Design** Conversations

Essential frameworks for killer design discussions

What this book is about

If you're a designer, design leader, or business leader, you probably have design conversations every day. Some go well. Some don't. Some drift, stall, or miss the mark. It's frustrating!

Do you want your design conversations to be more consistent, more focused, and more effective?

If you are a designer, leader or work in one of the roles that have evolved to encapsulate aspects of design, this book helps you get better outcomes more often. It brings structure to the conversations you're already having to support more meaningful business and customer impact.

It's not a rethinking of design practice. This book won't tell you how to design or which methods to use. Instead, we'll focus on how to communicate your decisions and reflect on your practice more clearly and effectively.

You'll find three visual frameworks to support different types of conversations, from strategic to day to day. Each is explained in depth and designed to be adapted. They're not strict formulas but flexible structures you can tailor to your context.

The frameworks are here to help teams stay aligned, reduce confusion, and focus on what matters. They show you how to cut the 'waffle' and zero in on what drives impact. We'll explore a set of core themes that should underpin most design conversations and that are worth tracking, repeating, and reporting on. Done well, this strengthens both your outcomes and your design practice's visibility.

There's nothing new here for the sake of being new. Much of it will feel familiar, and that's intentional. The value is in making what works more repeatable, more collaborative, and easier to build on. And to cut out the things that don't add much value to design conversations.

This book won't solve every problem, but it *will* help your team work better together. And that's where good design starts.

Table of contents

Chapter 1: Introduction

Cutting design 'waffle'

Business leaders often say that designers 'waffle on' when discussing design. We are passionate about what we do, so isn't it natural to want to share everything that goes into creating thoughtful, user-centred design?

Understandably, our audiences often don't know or care about these complexities. On the flip side, some designers simplify their message by focusing only on visuals, which can sell the design process short and leave audiences unaware of important design decisions. This risks good design being misunderstood or undone in the future.

The real challenge is that people's interests and knowledge about design vary widely. The solution isn't to 'waffle on' or reduce design down to visuals only. Instead, well-structured conversations that are flexible yet focused can make a world of difference.

This book aims to ease these frustrations by providing clear, adaptable frameworks for design conversations that work at every level within an organisation.

Impact of good and bad design conversations

The good: a win-win-win situation

Successful designers and design leaders often speak of the importance of quality design conversations. Those who work with them report enjoying discussing design and see the importance of truly improving their approach to design.

Important: Improving the approach to design within an organisation is a rare win-win-win situation for:

- designers who will feel more fulfilled in their work,
- organisations that can benefit from improved efficiency and revenue, and
- customers who have more enjoyable interactions.

Today, many designers report that they don't feel listened to and aren't having the impact they aspire to. A symptom is that many request a 'seat at the table' as they feel ignored and powerless to prioritise initiatives that they think will impact the customer and organisation. Meaningful design conversations ensure they feel heard.

Organisations are the big winners as they will be able to better identify efficiencies in workflows and increase revenues as they produce more attractive products and potentially identify more revenue streams.

Customers benefit from better interactions with the organisation through well-designed experience, improving interactions and removing annoyances. Hopefully, organisational efficiencies would also be passed on in part to the customer.

Most organisations say that they value design and put the customer at the heart of what they do. Most leaders say the same, and most people working in design feel that they are customer-centric. So, how come we continue to see poor quality design that doesn't fulfil customer needs?

Some studies have seen customer satisfaction with digital products steadily decline since 2018. Often, the commitment to user-centred design is just lip service, because without meaningful conversations, there is little action, or things get worse.

The bad: unseen negative impacts

Poor design communication can have a lot of seen and unseen negative impacts on a design practice. The most obvious is that stakeholders don't understand what is being spoken about and are not able to get on board. They can also mistakenly think the design practice shows a lack of progress, and they can either exclude design from discussions or put additional distance between themselves and design.

This is often achieved by having others speak on behalf of the design practice, so design loses its 'seat at the table'. This is observed in some organisations as reducing dedicated design leadership or demoting design under development, marketing, or product management structures.

Search LinkedIn or Medium and you will find the stories of design frustration. You will frequently see the same two factional solutions:

- Design needs to be more commercial, or
- organisations need to better understand design.

The reality is that organisations are already full of groups trying to better understand design and designers trying to better understand business. They know that by doing this, they will achieve their job goals. But there is often poor communication between the groups.

Design needs to lead a two-way conversation, where design practices are explained and there is an understanding of where design fits within organisational thinking. Assisting design practices are missing clear visual frameworks for design discussions. Business leaders generally appreciate frameworks as they give structure and focus, and frameworks themselves should not be underestimated as mechanisms for aligning knowledge and understanding.

Organisations without well-designed design conversations have frustrated stakeholders, teams, and design leaders that aren't fulfilling their potential. Some areas of particular frustration for design leaders are how colleagues consider design teams, both in terms of understanding what they produce and how they work.

Designers need to take responsibility for their own destiny

Successful design practices are made up of proactive designers and design leaders. They are proactively improving how they work and what they produce. They are also proactive in collaborating with others and leading conversations about design and customer centricity.

Important: Less successful design practices are often led by those who leave it to others to think about ways of working or expect to be involved without needing to state their case.

Proactivity is needed because, although many people in organisations value design, they don't necessarily understand it. In that void of design understanding, they will either be influenced by vested interests or continue in ignorance. Designers and design leaders should step into that void to lead clear conversations, shedding a light on design and customer centricity so that organisations can be more effective.

However, some design leaders have historically added to confusion when taking the lead in conversations by framing design as mysterious. For example, using terms like Design Ninja, Design Hero, and so on does nothing to help others support user-centred design practices.

Important: Showing a design process like Newman's squiggly line doesn't help business leaders understand when they will see results and how they can be involved.

Taking the lead in a conversation requires design leaders to direct the conversation away from low value topics but ensure that it is a two-way process. A conversation isn't just one side talking.

A seat at the table is not enough

High-quality design decisions and strategy can showcase the value designers can offer. Many designers request a 'seat at the table' to make critical decisions. This is understandable, as many report being excluded from key decision-making forums. They think that if they were able to call the shots, things would be better.

So, what's stopping them from having a 'seat at the table'? The harsh truth is that designers themselves are the reason designers aren't consulted as much as they'd like.

Important: Designers need to think about what value they bring to decision-making forums and make that added value clear. With that knowledge in stakeholders' hands, they will likely value design's input into discussions that they have been historically excluded from.

However, seats at the table can be quickly removed if the design team is not bringing value or isn't well understood. Having 'a seat at the table' is a useful by-product of a mature design practice, but many teams that have had this mythical seat have also had setbacks and failures.

Influencing the thinking that steers decisions through well-designed discussions is more important than being the decision maker.

Important: Designers need to avoid being design dictators who think they have the right to make all design decisions.

Centralised vs distributed vs fragmented design team conversations

Centralised design teams no doubt make it easier to measure an organisation's design maturity. However, the reality is that many organisations don't have centralised design teams. Design practices can be distributed, where they report to a single design leader, but day-to-day, they sit in another team. Designers are often fragmented, where they report to different leaders in different parts of the organisation or even in third party organisations. There are a variety of reasons for these different types of models and how they serve different organisational setups.

All the conversation frameworks in this book can be used no matter the design practice structure. However, different options for their use will be discussed later.

Facing the boom-and-bust cycle

Many design teams go through a boom-and-bust cycle, so it's important to be prepared. The organisation sees the theoretical value of design and hires a leader who then builds a large team. The value is then either not realised or there are budget crunches, so the team gets reduced or removed.

Regular, well-rounded conversations about design practices can help counteract this cycle in a few ways:

- They can make the team more efficient so that they don't grow to an unsustainable size.
- They can help highlight other opportunities to improve outcomes, for example, tools or strategy alignment.
- Most importantly, they can help the organisation understand the work the team does and the outcomes beyond the end product that the customer sees.

Bust cycles are often linked to a change in an organisational structure, for example, a level of management being removed or design reporting to a business leader without an appreciation of design practices. This is why an over-focus on 'having a seat at the table' can be a dangerous strategy.

A strong base for a design practice is the best safety net for any bust cycle. Ensuring basics such as workflow management, strong success measures, and strategic alignment will mean that the team will not fall as far during unforeseen organisational changes. Well-designed design conversations can help set these up.

The importance of conversation frameworks for design discussions

Many design leaders report struggling to have rounded conversations about how design and design practices can be improved. Those who are successful have well-structured conversations that achieve a lot of buy-in. Unfortunately, design education hasn't classically focused on how to have design conversations, especially at a management level.

Frameworks exist on what and how to design, but there aren't good frameworks about how to consistently evaluate and discuss a design practice, designers, and individual designs. This is not to say there aren't design practice maturity models, but many of these models are unrealistic and lack long-term value. In discussing design practices, people often just fall back on individual success metrics, or worse, 'best practice'.

Important: Success metrics are important for understanding impact level but not how you can increase that impact.

The design conversation frameworks outlined in this book are intended to work alongside high-level success metrics to bridge that gap.

Using success measures

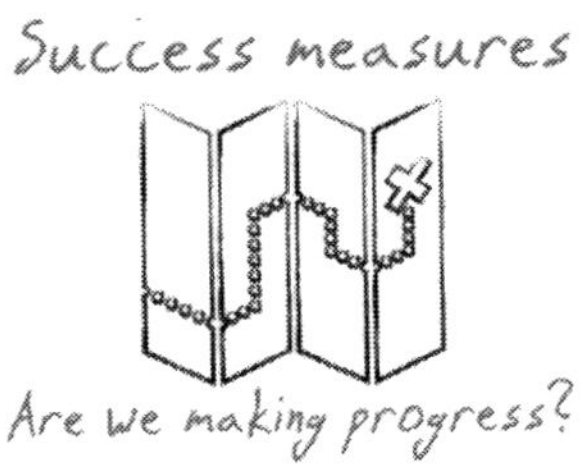

Frameworks go hand in hand with success measures, which are important to aid in tracking the impact of any changes. Conversation frameworks help you understand what topics or categories you might want to focus on, while success measures help you understand the impact of focusing on the categories.

How success is measured varies between different organisations and teams. Many articles discuss different options for success measures for design practices, designers, and designs. There has been a lot of great thinking in this area, which I encourage you to explore. Most focus on customer and business impact and team and stakeholder satisfaction.

One area that doesn't get much coverage but is also worth considering for both design practice and designer reviews is the organisational belief in the design practice or designer to make a difference. Stakeholders believing that the design practice or designer can have an impact on both the customer and organisation helps identify where focus is needed. For example, belief in the potential impact of design practice is low could mean more focus on strategic alignment is required. A high belief may indicate a need to focus on design craft skills to maximise efficiency.

Important: With any success measure, it is vital to:

- find out what others in the organisation value most,
- measure against that, and
- share those measures openly.

Using visuals to support design conversations

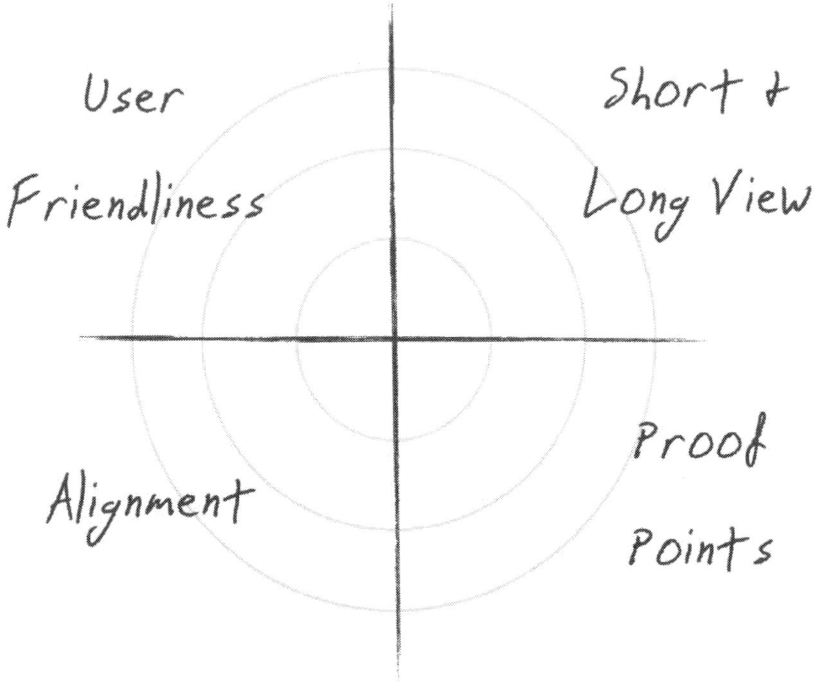

Visuals can assist conversations by providing information in a holistic way so that trade-offs can be seen, viewers can quickly get up to speed, and questions can be asked about categories of interest. Imagine talking about yearly financial results without a graph. It is possible, but it makes any message harder to deliver.

Important: Visuals assist design wafflers to include important information but without spending more time than necessary.

They assist design leaders in showing design output and the thinking behind it, together with trade-offs, without losing the audience.

For example, the double diamond design process is arguably the most influential design process diagram to ever exist. Created by the UK Design Council and popularised by IDEO, the double diamond is a simple visualisation of a four-step process, from ideation through to design delivery. Love it or hate it, double diamond has had more impact on organisational understanding of design than any other concept. It has a lot of complex thinking behind it, but it is quick and simple for anyone to pick up. Why? It has a catchy name, it has four simple focused steps, and it has an easy-to-replicate visual. It is an impressive piece of communication design that has been a great aid to many discussing design processes.

Designers are well equipped to use visuals to support conversations, so simple visuals have been used to bring to life the three frameworks in this book. You should be able to easily replicate these in online and offline whiteboards, presentations, and more.

Well-designed design conversations are not one-off events

Design conversations are revisited and revised constantly. This is particularly important when there are changes in the team and the environment in which they operate. Organisational changes can have potentially positive and/or negative impacts on design practices and what is delivered to customers. Frameworks that are easy to share with stakeholders can make it easier to reduce any negative impacts of organisational changes outside of the design practice's control.

Important: Design conversations must be realistic, meaning not just theoretical, where anything is possible, but realistic, where trade-offs need to be discussed.

Every day, designers, leaders, and others in the organisation have trade-off conversations about what is wanted and what is possible within the limits of individual initiatives. The conversation frameworks in this book can be used to visualise these trade-offs.

Design practices, designers, and designs evolve over time. Design conversations are revisited as changes are made. However, that history can be lost or unknown to those coming to conversations later on. The design conversation frameworks in this book can be used to show changes over time.

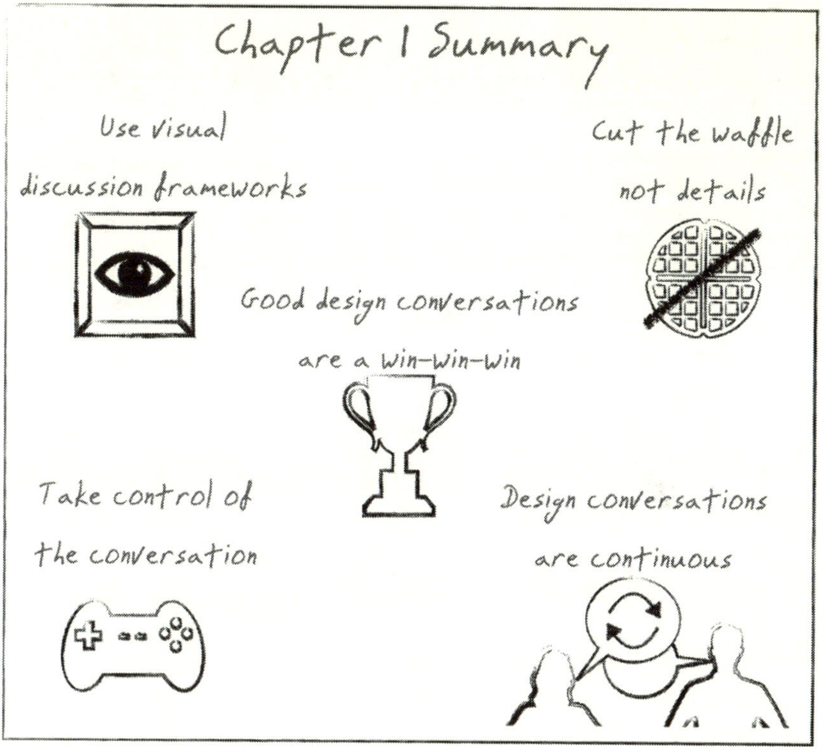
Chapter 1 Summary
Use visual discussion frameworks
Cut the waffle not details
Good design conversations are a win-win-win
Take control of the conversation
Design conversations are continuous

Chapter 2: The three design conversations to target first

This book puts forward three design conversation frameworks that support structured discussions about design and design practices. They are:

1. **The Design Practice Conversation Framework:** covering the design practice as a whole,
2. **The Designer Conversation Framework**: covering the designers who make up the practice, and
3. **The Design Conversation Framework**: covering the designs created by those designers.

These frameworks are designed to be easily shared with teams and stakeholders. They ensure that the most important factors are targeted during conversations. These rounded factors about successful designs are not just the design's visual appeal but also its potential impact and commercial sensitivities.

You might find that design conversations need to flex depending on the audience and their interests. That's why the frameworks are visual and flexible, allowing the audience to see all of the factors the designer or design leader is considering, but choose which area they want to cover in depth.

The design practice conversation framework

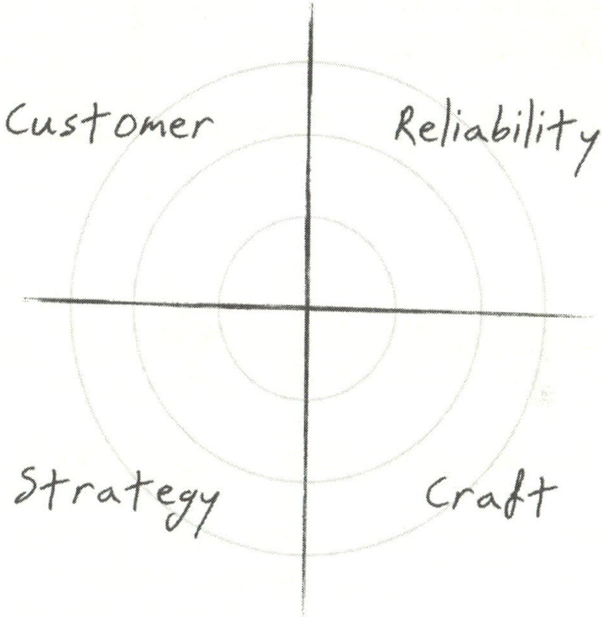

Busy workplaces always have stakeholders to meet with, initiatives to action, designs to review, and team members to check in on. For design leaders, these activities often take up a lot of the day and can be used as a shield to avoid deeper thinking about the overall practice and creation of impactful strategies. It is easy to hide behind 'deliverables', and many design leaders have made a career out of this. They deliver what has been asked for and are seen as a safe pair of hands. However, delivering the design basics and not improving the organisation's user-centred culture is also something AI is very good at too.

For those design leaders who do carve out time to look at their design practice and ways to improve, it can be a challenge to know where to start discussions with the team and stakeholders. This is where the design practice conversation framework is a useful tool for planning design practice conversations. It provides a framework for this thinking and enables progress tracking and for these improvement areas to be shared.

Why not just use the existing maturity models for assessing a design practice?

Important: This conversation framework is different from many existing maturity models because it shines a light on strategic focus and repeatability, both of which help ensure that the design practice is successful now and into the future.

Maturity models can be powerful tools throughout a design leader's career. They are worth reviewing to understand the thinking behind them. However, many issues have seen them fall out of favour in recent years. There seem to be a few reasons for this:

- Teams taking a purely project/initiative-focused approach
- Lack of awareness and a feeling that they are for 'newbies'
- The impracticality of some of the well-known models

The design practice conversation framework in this book aims to offer continued value to design leaders as it helps them track progress.

Important: The framework in this book is designed to be flexible, so if some of the categories don't fit the organisation's needs at the time, they don't have to be addressed.

Most existing maturity models are complex, use jargon, or are theoretical rather than applicable in the real world. Having a practical model that is easy to use, share, and revisit is vital. The ability to quickly and reliably map a team against a model enables leaders to quickly grasp opportunities and strategies. Being easy to share means it is quick to sketch and easy to discuss.

Many maturity models have several levels that must be completed to move up to the next level of maturity. Often, the highest levels of design maturity models focus on having 'a voice at the executive table'. As mentioned earlier, this is not a good indicator of design practice maturity. For example, chief design or customer officer roles are created to give design a voice but are often filled with people from business backgrounds rather than those who possess design sensibilities.

One model that took a different approach and inspired some of the frameworks in this section is the Muzli Business Design Maturity Model. The strength of this model was to recognise that organisations often don't have a linear approach to building design maturity. One organisation can be very strong in one area and weak in another. This is important as it means there isn't a cookie-cutter approach to improving design capabilities, but similar characteristics should be considered. Another useful aspect of this model is that it has a self-assessment survey that anyone can use to review a design capability. Although similarities are shown between their model and the one in this section, they are fundamentally quite different.

Practices can regress

An area often not considered is maturity regression. Many organisations have made great progress with maturing their design practices only to then regress. A step backwards can have many causes, such as:

- new strategic direction,
- new organisational models,
- personnel changes, and
- budget pressures.

Important: The fact of the matter is that most design practices will regress at some point.

With this in mind, the design practice conversation framework attempts to soften any backward steps, so teams are ready to progress again in the future.

Design is just one consideration

Lastly, design practices often struggle as they focus just on how design is created. This is an insular view rather than the reality of many organisations where the design function is just one consideration amongst many others.

The individual designer conversation framework

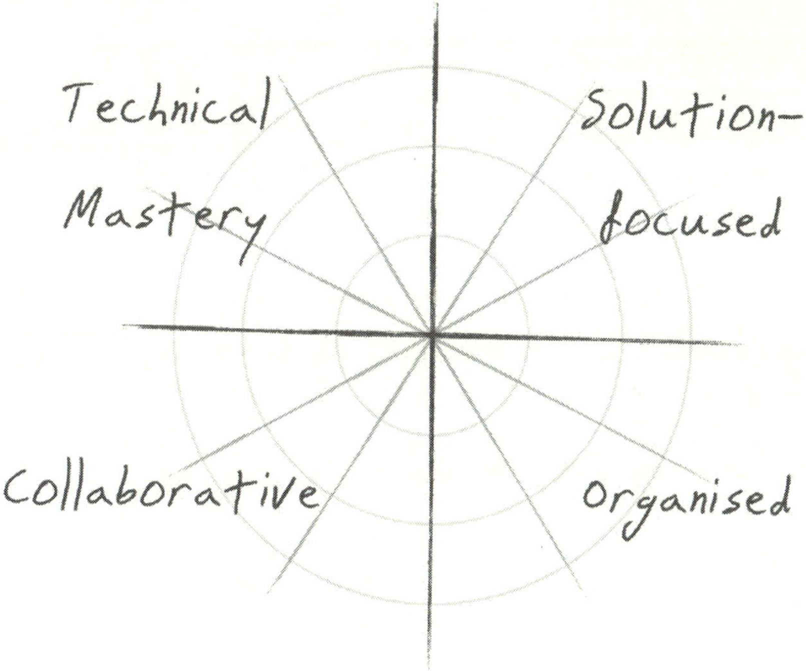

Skilled designers are at the heart of successful design practices. Designers can take many forms, so this section uses the term in a general sense to cover UI designers, UX designers, product designers, service designers, and so on. From speaking to leaders and reviewing thousands of job specifications, it is clear there is a gap between what is required and what is requested. The framework should be used for assessing the designer's qualities. If a designer possesses the discussed qualities, they are more likely to produce strong work and assist improving the overall design practice.

The individual designer conversation framework is aimed at supporting professional development conversations between a designer and their manager. It consistently shows where the designer should be focusing their professional development and any progress made. The framework enables open and honest conversations so that the designer and their manager can share where they both feel the designer has strengths and weaknesses. If there are differences between what the designer and manager feels, then the framework highlights these for discussion.

Speak with older workers with many decades of experience and you are likely to hear a repeating theme that keeps them motivated: They enjoy always learning new things in their roles. No one will ever master every element of a role. As the saying goes: 'The more you know, the more you know you don't know'.

Important: Workplaces, technologies, and understanding are constantly evolving, so standing still is falling behind for designers.

Being open and honest builds trust

Having a clear, consistent framework for conversations with designers about areas of strength and opportunity sets clear expectations. It also lets the designer know that the same open and honest conversations are happening with other designers in the team.

Important: The trust that all members of the team are having similar conversations not only improves the relationships within the team but also means it is more likely that the designers will work together to develop skills.

Using the framework for hiring

This conversation framework can also be used when hiring new designers. Use it to think about which of the qualities is most important to your team and look for people who demonstrate this.

But why not just use job descriptions to assess a designer? The individual designer conversation framework presented in this book is intended as a way of structuring discussions with designers about professional development rather than an official employment review document. Also, the nature and quality of job descriptions for designers vary substantially between organisations. The conversation framework for individual designers in this book can be used to form a section of a job description. However, job descriptions are often more complex and include organisation-specific expectations, such as reporting lines and success measures.

During interviews, the most popular approach is just to ask potential designers for examples of projects and score them on how they dealt with them. The challenge is that this often results in a simplistic answer with little depth of understanding of how they would deal with future situations. Sharing the designer conversation framework with the designer and asking them to speak to their strengths and weaknesses enables them to speak to specifics. Saying they are strong in all areas is a warning sign, as often the more experience someone has in a category the more they will know their limitations. Designers who identify areas they want to strengthen can be a sign of maturity.

The design conversation framework

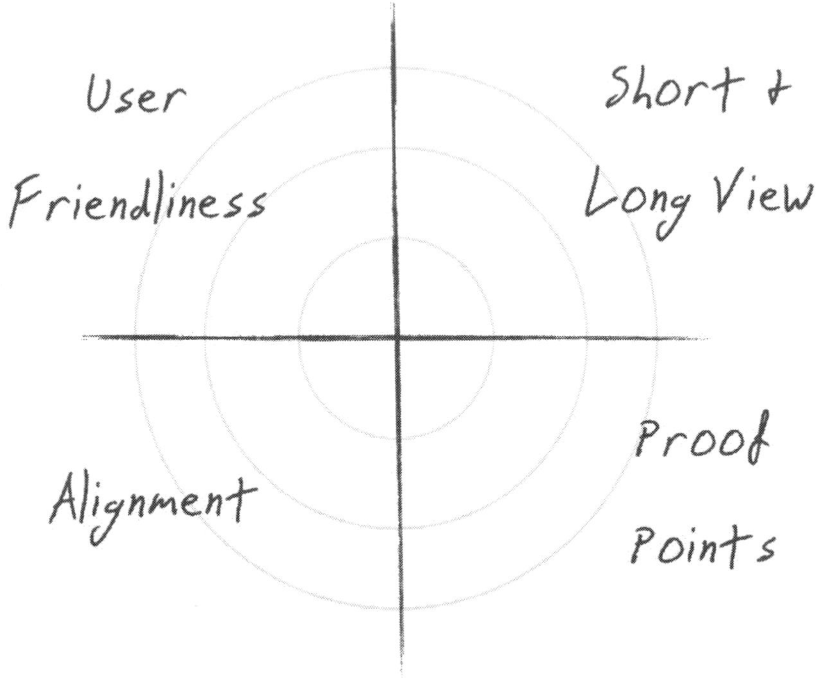

Everyone has a view about design. It's an area that almost everyone in an organisation feels they can add value, whether or not they're an expert. You have probably met designers frustrated by the influx of feedback they receive about their designs. The people giving the feedback are often unaware of the trade-offs the designer makes. So, giving reviewers more context will improve feedback quality and help set expectations.

This framework aims to be usable for both designers and reviewers. It visualises design considerations as well as strengths and weaknesses. It is generalised to work across all types of design. A great design doesn't need to have a top score in all areas, as there are often trade-offs. For example, some useful features might not be feasible in the budget of your initiative. This model can be updated as the design changes throughout the design and delivery process, which can be useful for discussions about unforeseen opportunities.

Why 'best practice' isn't always the best fit

Why not just use 'best practice' to assess design quality? Best practice is one of the most widely-used and meaningless terms in design. It rarely is intended to encompass international design standards such as ISO 9241, ISO 18529:2000, and WCAG 2.2, as many teams are either not applying these or are unaware of them.

Over the years, the more the term 'best practice' is used, the less effective it becomes. Often, when designers refer to 'best practice' it is purely opinion-based, has very limited proof behind it, or is a misunderstanding of psychological concepts. For example, a large number of designers believe that Miller's 7+/-2 law is based on sound science and applicable to elements such as menu design when it's neither of these things.

The muddy nature of views on design 'best practice' can take focus away from a rounded discussion of customer needs and where compromises can be made.

Important: It may lessen the belief in a designer's skills if they are not able to clearly articulate the benefits of their design decisions beyond 'it's best practice'.

Staying on the same page as designs evolve

Using a design conversation framework whenever designs are reviewed ensures everyone is on the same page regarding the most important factors throughout the design discussion.

The design conversation framework enables consistency when discussing designs throughout the design and delivery process. It ensures that factors beyond the design's ease of use are covered and enables stakeholders to understand design considerations.

Important: The framework also visualises the trade-offs that happen with every design.

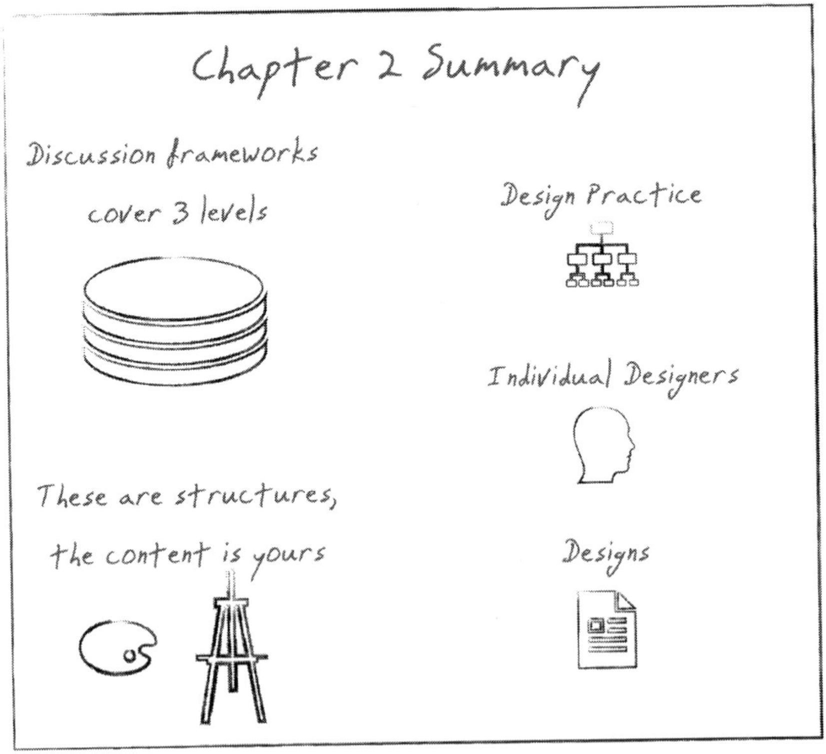

Chapter 3: Using the conversation frameworks

Before sharing more details of the conversation frameworks, it's important to outline how they are intended to be used. The frameworks are designed to be quickly understood and applied. They are also extensible, so that if needed, a lot more depth can be explored. While they are designed to be used by design leaders, they are also easy to use for organisations without dedicated design leadership roles.

Important: Whether being used quickly or within an in-depth review, it is vital that evidence, not a vibe check, is used to complete the frameworks. This leads to a higher level of confidence in the reviews and avoids any negative thoughts of a design practice not being aware of its shortcomings.

Remember: The aim of using the frameworks isn't to reach perfection.

Wrongly or rightly, designers are often seen as perfectionists. However, I've heard many designers reject this sentiment and complain that their pragmatic design choices aren't recognised. Ultimately, trade-offs will always need to be made. The frameworks enable you to visualise and discuss them.

Filling in scores

The frameworks show the maturity level of the design practice, designer, or designs to prompt conversations. The scoring highlights which categories (and sub-categories) may need attention. There are two approaches you can take based on your needs:

- Approach 1: The simplest approach is to score using the main four categories only.
- Approach 2: The more in-depth approach is to score using the three sub-categories that each category contains.

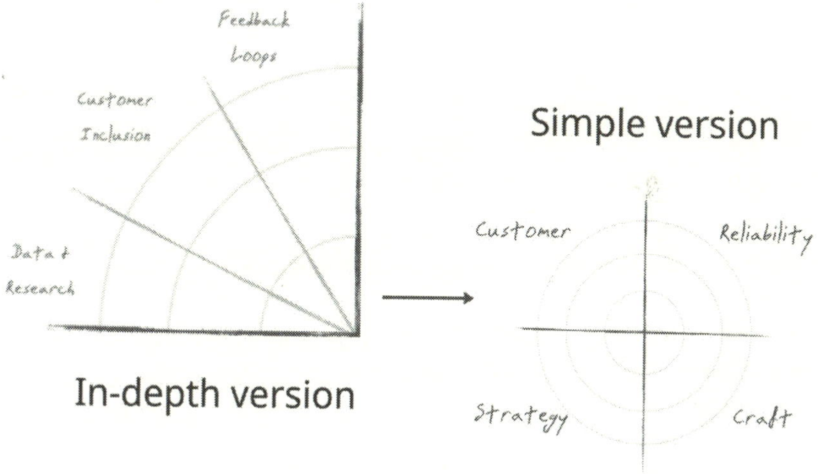

Scores are based on the visible activities of the design practice, designer, or design. These visible activities are called 'tell-tale signs'. If these tell-tale signs are not seen, it is unlikely the design practice or designer is acting at that level. This avoids subjectiveness.

With conversation frameworks 1 and 2, design practice and designer conversation frameworks, it is also possible to run surveys or interviews with stakeholders to understand how they would score the design practice/designer. This adds an extra dimension to the conversations as it helps you understand if the design practice/designer's perception is aligned with the evidence-based reality.

Important: A common challenge with the individual designer conversation framework is that a designer is likely to overscore themselves. In this situation, it is important to refer to evidence and discuss why there might be a difference between the self-view and the evidence-driven view.

Standardising scores

A challenge with these types of frameworks is that different people have different ways of scoring design practice, designer, or designs, even when using the same categories.

To move away from arbitrary opinion or vibe-based scoring, the frameworks presented here require evidence to achieve scores. Without something tangible to point to, a design practice will not be scored highly in an area. For example, a design leader may feel that their design practice is customer-focused, but without clear evidence, such as customer feedback loops, the 'Customer Focus' category shouldn't be scored highly.

Gathering evidence-based tell-tale signs

Marking against the three frameworks in the following chapters requires gathering evidence. Each category has a list of tell-tale signs indicating whether the design practice, designer, or design (depending on which conversation framework is used) should score highly or not. It should be possible to point towards evidence for each of these tell-tale signs. That evidence could be a document, research, analytics, conversation, process, or feedback. For example, to score well in the 'Alignment': 'Feasibility' sub-category of the design conversation framework, a designer should be able to reference conversations with others where they have reviewed the design from a feasibility perspective. The designer can't just determine that their design should be feasible.

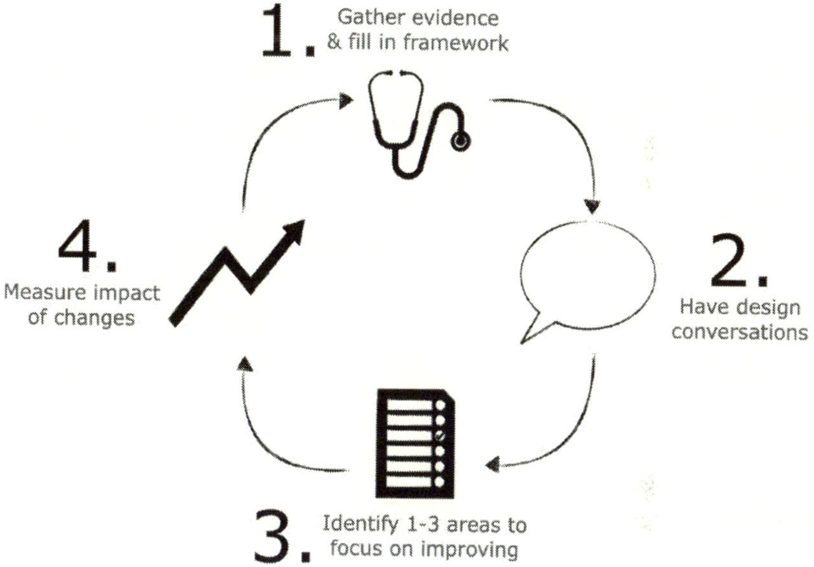

Approaching conversations in complex structures

Many organisations have fragmented or distributed design practices, either where there isn't a single design lead or where several co-exist in different parts of the organisation's structure. In these situations, consider the following approaches:

1. Review each design team individually.
2. Review the design practice holistically.

Both have potential considerations that need to be accounted for.

1. Reviewing each design team fragment individually

Issues with overlapping success measures can occur if each fragmented design team is reviewed individually. For example, if two design teams both produce work for the same customer group, measuring the impact of just one of the teams on the customer can be difficult.

This is where targeted measures may be needed to measure the impact of the customer journey part that each team contributes to. Analytics and A/B testing are useful tools to consider in this situation.

2. Reviewing the fragmented design practice holistically

If reviewing the fragmented design practice as a whole, the practice may score highly because of one design team's approach while hiding areas for improvement in another.

In this case, it may be worth giving a range score to show that the good practices are not consistent across the organisation.

Putting your conversation frameworks into action

The frameworks in this book are a cookie-cutter approach for easy adoption but are broad enough in scope to be used flexibly.

While they should be used as the framework for conversations, the substance of the conversations will vary depending on organisational needs, designer seniority, and type of design under review.

Remember: The frameworks are a starting point for structuring design conversations, but the content of those conversations will vary.

Once you have scored the design practice, designer, or design against the relevant framework, you are likely to see areas of strength and weakness. Just because an area is strong it doesn't mean it should be neglected in the future. A category identified as important is likely to require ongoing focus, even when a high level is reached.

Strengthening areas of weakness

When improving areas of weakness, it is important to examine any trade-offs that may have to be made. For example, there may be a budget or alignment trade-off to uplift a current area of design weakness.

Important: Try not to take on too many categories at one time so that the design practice doesn't feel over-stretched or lose clarity.

There can be many reasons for weakness in any particular area. Some common barriers to maturity have been listed in later chapters, along with strategies that have worked for others in the past.

The road ahead

The completed conversation frameworks should be kept up to date and shared again whenever the topic arises. Sharing the scored framework and next steps with stakeholders is crucial to encourage deeper conversations.

Remember: The completed conversation frameworks should be updated to reflect changes and shared whenever the topic is discussed to aid in effective design discussions' planning and delivery.
Showing where changes have been made based on historical conversations helps reassure stakeholders that those historical conversations had an impact. For example, if the designer agrees that more effort is required to ensure their design's feasibility, the design conversation framework can be updated to show the change in this feasibility focus. Examples of this are shown in the worked example sections later in the book.

Flexible language

Simplified language has been used to improve the three frameworks' accessibility and universality. Your organisation may have different terminology, or you may want to use terms currently popular in the design community. You can easily swap out the terminology in the frameworks, but try to avoid niche terms when more universal options are available.

Beyond the conversation frameworks

The conversation frameworks in this book do not cover aspects of design conversations such as tone, timing, and style.

Important: The conversation frameworks will support well-structured design conversations. They should be coupled with designers and design leaders building their own presentation and communication skills.

Storytelling will suit some, and having a stack of stats to back up what's being discussed will suit others.

It's particularly important to choose the right forum. Will it be in person, and will it leverage existing forums? Spend a good amount of time thinking about how the design will be delivered, as well as continually using the frameworks to keep on track. Later in the book, we will revisit the topic of where and how to have these design conversations for the most impact.

Using consultancies

Using external consultancies can be a good option for organisations that:

- do not have a reliable design leader in place,
- are looking for an independent review, or
- want an in-depth review but don't have the capacity to take this on.

The frameworks in this book should be easy to pick up and use by any competent design consultancy.

A closer look at the three conversation frameworks

In the coming chapters, each of the three frameworks will be looked at in depth, including what makes up the frameworks, tell-tale signs used to score against them, common barriers to scoring well, and worked examples to see them in action.

Now, let's dive in!

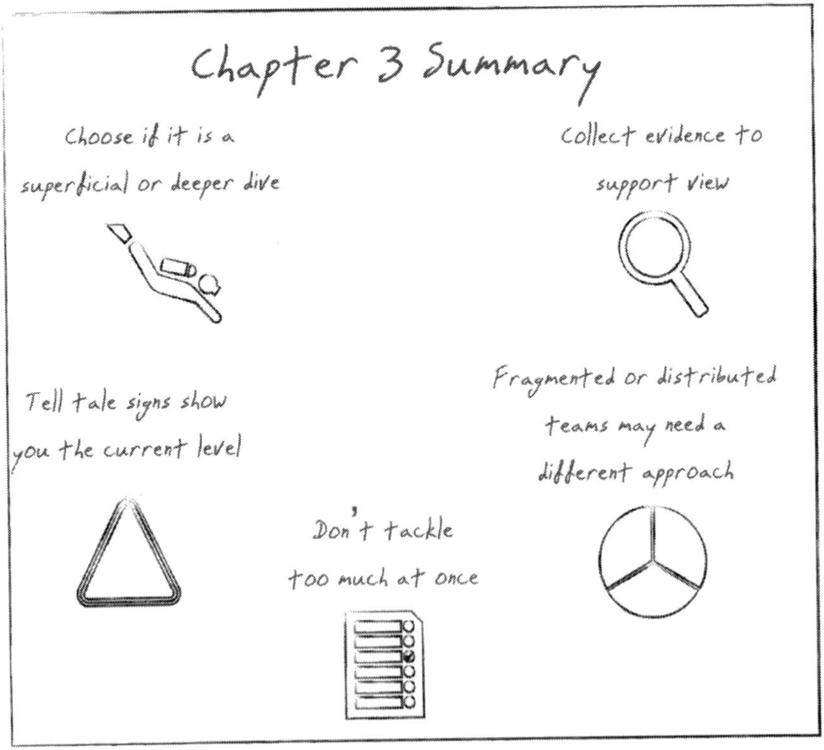

Chapter 4: Applying the design practice conversation framework

Balancing practice improvements with ongoing design delivery work is important; the world won't stop just because a design practice wants to improve its approach.

Remember: This framework doesn't suggest that a design practice must reach the highest points in each category. It simply highlights opportunities to improve if that category is a priority. If a design practice scores low in many categories, it is recommended that only one or two are tackled at any time, so the practice isn't overly stretched.

The design practice conversation framework categories are limited to focus on areas that deliver the most value to organisations. When planning design practice conversations, it is important to have a limited number of clear messages and focus areas. This ensures stakeholders and the team have clarity over the most valuable opportunities at hand.

Important: Avoid the temptation to add more categories, as this will lessen conversation impact.

Setting strong foundations is particularly important during an organisational change, for example, resource crunches or rapid strategy changes. Without strong foundations, reduced-strength teams can struggle or sometimes be wiped out.

The four framework categories are:

- Customer
- Strategy
- Craft
- Reliability

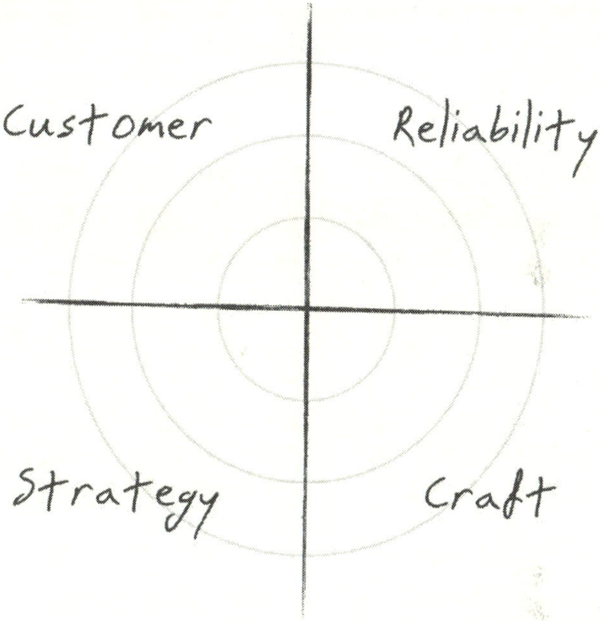

Category: Customer

Being customer-focused is important for all teams, but it's vital for a design team. You can't design without understanding customers. Design provides an appealing and useful output, meaning that any design's success lies with the customer or user. A design team without a focus on customers has a higher likelihood of creating something the customer doesn't respond to or care about, wasting resources.

Being customer-focused is rarely challenged in the modern workplace. However, saying it and being it can be two different things. The proof is the actions and processes that are consistently observed.

The sub-categories for this section help shine a light on opportunities for the design practice to take the initiative with customer centricity.

The three key sub-categories of a customer focus are:

- Data & Research
- Customer Inclusion
- Feedback Loops

Customer

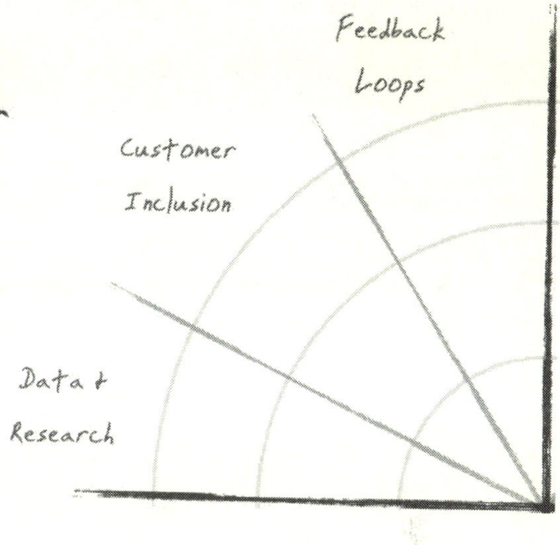

Customer: Data & Research

One of the traits that has made humans so successful is our ability to observe and adapt. It's helped us hunt, farm, and form deep relationships. However, we also tend to act on feelings and biases in order to move quickly. Without these traits, our achievements so far wouldn't be possible. But we wouldn't thrive without a balance. Designers naturally rely on feelings and biases because it's the path of least resistance. Using data and insights to observe and adapt takes a bit more conscious effort.

Gather iterative insights

Ideas and designs evolve throughout the design and delivery process. That's why it's so important to gather insights iteratively. Just gathering this information at the start of a project will not show you if your solution has hit the mark. Just gathering data and insights at the end of the delivery process will increase the risk of expensive reworking or lead to missed opportunities. An iterative approach to gathering insights avoids these issues.

Important: Mature practices will have a lean approach that works within budgets but with the rigour required to ensure that the data and insights can be relied on.

Communicate insights

Effectively communicating insights with the team and stakeholders is a win-win. It ensures better customer outcomes while speeding up design approvals. This requires the team to focus on the important information and present it in an easy-to-digest way. Design teams should use their existing skills to design data and insight reporting with the internal consumer in mind.

Teams that aid their stakeholders in better understanding customers are typically more impactful. They have to spend less time justifying ideas as stakeholders will know the customer context. Stakeholders are also less likely to push low impact ideas.

Your customers aren't just looking at you

Customers are aware of your competitors. In some cases, this helps shape their expectations of your product. Stakeholders may be even more likely to have awareness of competitor design solutions, as many design teams report being pointed to competitors for inspiration.

Understanding a competitor's good and bad aspects is vital for adopting only the best ideas. This helps teams explain to stakeholders where the differences and similarities lie in designs. Just as understanding your own product is important, customer feedback and insights are critical to understanding competitors.

Tools such as AI agents make gathering competitor insights and coming across their updates much easier than ever before.

Important: Effectively and regularly communicating competitor insights with the team and stakeholders ensures the team is respected for knowing the competitive landscape in which they operate.

Tell-tale signs of a mature design practice

- ☐ Design plans include data/insights tasks.
- ☐ User/customer insights are regularly shared.
- ☐ Competitor experience analysis is frequently conducted and shared.
- ☐ Late-stage design changes are infrequent unless new customer information comes to light.
- ☐ Design reviews include data that justifies design decisions.

Potential barriers to scoring well

Data/research are limited.

Tactics to consider:

- Partner with other teams who have access to useful data.
- Set up training and processes so that designers can gather data/insights when needed.
- Request investment in customer research and analytic programmes to feed into projects.

Lack of access to existing data/insights.

Tactics to consider:

- Build stronger partnerships with teams who can grant access to existing data/insights.
- Escalate to leadership to gain access.

Lack of resources to run research.

Tactics to consider:

- Develop an investment business case for research.
- Look for low-cost and high-speed research options.
- Create case studies where research has had positive commercial impacts, and share with leadership.

Research is seen as slowing down the design and delivery process.

Tactics to consider:

- Look for opportunities to run case studies of rapid/lean research to demonstrate how it can fit into project timelines.
- Show timelines and costs for different research approaches.

Customer: Customer Inclusion

Including the customer in the design process is a tell-tale mark of whether customer centricity is window dressing or truly in a design team's mindset.

There are many ways to include customers in the design process. Detractors of customer inclusion will often use the Henry Ford quote, 'If I'd asked customers what they wanted, they would have told me, "A faster horse!"'. There is no evidence Henry Ford said this, and it misses the point of customer inclusion.

Important: It's not about just acting on what customers ask for; it's about understanding why they are asking for it and ensuring it is front of mind at every stage of the design process.

Avoid designing for yourself

'Designing for self' is a challenge many designers face. Often, employees believe they know the customer inside and out because they, too, are the customer. But they may only represent a very small segment of the customer base, while their level of knowledge and relationship with the organisation further sets them apart. There can also be a wide gulf in education and circumstances between the average employee and their average customer.

Avoid designing for a customer only you understand

Stakeholders and designers often reference a generic 'customer' as a substitute for personal preferences. This creates the impression that the customer is the focus, but it can lead to misunderstandings.

Important: One person's view of the customer may differ from how others picture them in their mind. Without a shared understanding of who the customer is, it's difficult to build empathy and align on their needs, resulting in more top-down, 'captain's call' decisions.

Having a clear, shareable view of the customer improves the quality and consistency of design decisions.

Even when a design team has a clear picture of the target customer, different business areas across the organisation may have fragmented (or conflicting) views of who the customer is. This is sometimes useful as different teams focus on subsegments of customers. For example, customer service focuses on customers with additional support while a sales team will focus on all potential customers. That's why it's important to have a consistent approach to describing the different customer groups so that they are understood and conflicting priorities are avoided during the design process.

Including the customer in the design process

Tools such as Jobs To Be Done (JTBD), personas, and segmentation have all been used successfully, when executed well. However, there are also many examples of these falling short when poorly executed.

Important: This doesn't necessarily mean that the tool isn't effective. Often, the issue is poor execution rather than the tool itself.

Using these tools can help, but even if a tool isn't used, the customer should always be considered when discussing design. Otherwise, it's a nasty shock when the customer judges what you've delivered.

Tell-tale signs of a mature design practice

- ☐ Design reviews always include information about who the design is for and how their needs are met.
- ☐ Designers consistently have meaningful conversations about how user needs are, or aren't, met.
- ☐ Stakeholders are aware of the target user-groups.
- ☐ Tools such as personas, segmentation, and JTBD frameworks are utilised consistently.
- ☐ Design leaders review and assess work quality based on meeting customer needs.

Potential barriers to scoring well

Organisational resistance to focusing on particular customer groups.

Tactics to consider:

- Work with senior leadership to gain an understanding of how they define the target customer.
- Partner with other areas, such as marketing, to define target customers.
- Highlight which types of customers are and aren't served well by any design solution.

Teams lose sight of the customer later in the design and delivery process.

Tactics to consider:

- Work customer references into session guides for design and delivery reviews.
- Publicise target customer groups internally, so they are better understood and impossible to ignore.

Designers are not comfortable talking about how their designs impact customers.

Tactics to consider:

- Create reusable artifacts to support designers.
- Run training for designers.
- Question customer impact in every design review.

No good understanding of customers and how they interact with the service.

Tactics to consider:

- Leverage insights gathered by other teams, such as marketing.
- Undertake research to gain an understanding of customers.

Customer: Feedback Loops

When do you know if a design is successful? Only when it's out in the world in the user's hands. Before that, you may have a hypothesis based on research, feedback from delivery teams, and projections from business teams.

Ideally, customer feedback avenues will already exist in the organisation. In this case, the design team must make sure they have access to it. However, customer feedback loops sometimes don't exist or are not fit for the team's purposes. When this is the case, the team needs to proactively set up these feedback loops or work with other teams that can help establish them. An immature team will throw their hands in the air and say they can't do anything about it. Mature teams will be hungry for feedback and so are proactive in collecting feedback and acting on it.

Feedback should trigger action

Another situation is when feedback is gathered but ignored. A mature team will not only gather feedback but also work with other teams to design solutions that resolve issues. However, even mature teams can struggle to get their improvements prioritised. This is often when they feel the need for a 'seat at the table' to get improvements actioned on backlogs. Although having a seat at the prioritisation table can help, it still requires a step that immature design practices often overlook: developing solid business cases that are easy to compare against other delivery backlog items.

Work with the organisation to prioritise action

Actioning feedback requires the team to work with business leaders, clearly articulate the ROI of acting on the feedback, and report the results.

Important: The team must also size the impact of the feedback. Is it isolated or affecting a number of valuable customers? Not every item of feedback carries the same value.

Tell-tale signs of a mature design practice

- ☐ Designers actively seek out customer feedback from other teams, for example, customer service.
- ☐ Journey maps/service blueprints include information about how to get feedback from customers.
- ☐ There is a roadmap for setting up or improving feedback collection.
- ☐ Delivery/product backlogs contain items created or championed by the design team.
- ☐ Insights on the impact of new designs are regularly shared.

Potential barriers to scoring well

Moving on to the next initiative without looking back.

Tactics to consider:

- Work with your team to add customer feedback reviews to every initiative.
- Create a reusable template for quickly publishing post-release insights.
- Work with project/product managers to ensure initiatives are not marked as 'done' until a customer feedback review is complete.

Not having an iterative delivery approach, leaving no room for action.

Tactics to consider:

- Work with product/project managers to understand how follow-up work is prioritised.
- Build up shareable case studies where iterative design improvements resulted in positive ROI.

Not having the ability to add to product backlogs.

Tactics to consider:

- Escalate with leadership to secure an avenue for carrying out design initiatives.

Wanting to look successful and avoiding any potentially negative feedback.

Tactics to consider:

- Change the design practice mindset to an experimental one that embraces learning from failures.
- Incentivise designers to address feedback issues.

Category: Strategy

One of the clearest differences between an average and a mature design practice will be the team's ability to create and contribute to organisational strategies. This enables them to go beyond impacting individual projects to impacting broader outcomes, including ways of working. This leads to designers being held in higher esteem within the organisation.

Design strategies often focus on change, both in what is being worked on but also ways of working. This can sometimes put the team at odds with other business areas that are resistant to change. They may have strategies focused on doing more but not necessarily on doing things differently.

Good design strategies are usually data-driven, which can lead to conflict with other business areas that are opinion-driven or focused on different data sources.

Unfortunately, many design and design leadership courses don't cover strategy creation mechanics. However, a lot of literature on the different approaches to strategy creation can be used.

In a gap of strategic capabilities, organisations will often outsource this to management consultants. This can be a risky tactic as few management consultants have a sound understand of design strategy. Making it even more important that design leaders learn how to create design strategies that are fit for purpose and ensure they are well communicated with senior leadership.

Important: If a design practice earns senior leadership trust, they are likely to be given more control of the strategies that impact their ways of working.

The three key sub-categories of the strategy focus are:

- Alignment
- How & What
- Investment

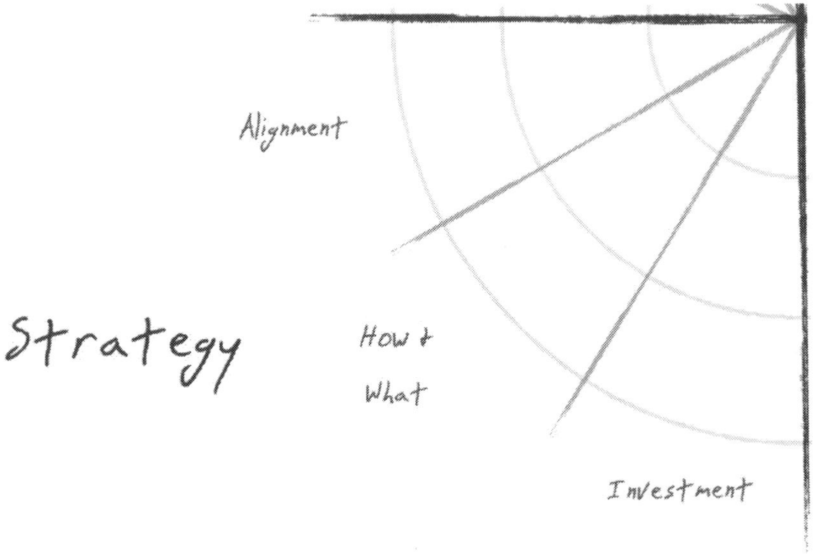

Strategy: Alignment

Like society, organisations are made up of many groups with different tactics and mindsets. But unlike society, there's no regular vote on which tactics are better than others. However, there's usually more clarity on the objectives for each group. This comes in the form of strategies and sub-strategies, making alignment easier.

Design strategies 'click' when everyone is pulling in the same direction. A mature design practice is aligned within the team, with partner teams, and with the organisation's objectives.

Important: Aligning design strategies with other teams helps build trust in the design practice and often leads to more freedom, as there's trust that the design team won't accidentally implement activities that negatively impact other areas.

The team has to be involved, not just told

There are many reasons why a team might not be strategically aligned. Often, it's because they haven't been involved in strategy creation or they've been unable to ask questions about it. This top-down approach increases the risk of the strategy missing key opportunities, while alienated team members are less likely to advocate for it behind the scenes.

Ensuring the whole team understands the design strategy reduces output and ways of working issues that threaten its delivery.

Strategies should converge

Partner teams such as development, product, brand, marketing, and operations will often have their own formal or informal strategies. Again, alignment with these teams is critical. The design team should input into these strategies. If not, they must fully understand existing strategies to see where they can leverage them or identify opportunities to enhance them.

Important: It's vital that partner teams without existing strategies input into the design strategy to help ensure its feasibility and increase the likelihood of senior leadership adoption.

A frequent challenge for a design practice is that they have to balance conflicting strategies between stakeholders. For example, one team may focus its strategy on improving efficiency in an area while another may prioritise sales at any cost. If the design strategy is pulled in two directions, then it is critical to be evidence-led, transparent, and provide options. This way, stakeholders can understand the different directions the team could follow and make an informed decision on which strategy to align design with.

Avoid cookie-cutters

Design leaders will sometimes take a cookie-cutter approach to their strategy, using what they have seen work well in other organisations. Although many organisations say they 'value design', what this looks like can be quite different. Design teams need to be aware of the organisation's strategy and how to ensure outcomes are aligned. For example, an organisation focused on reducing costs requires a different team approach to one focused purely on growth.

One or two-way doors?

Important: Any design strategy should highlight which decisions are one or two-way, i.e. which decisions are easily changed later and which ones are much harder to change.

Aligning with other teams helps them understand the potential impact of the design practice approach, together with their own one-way decisions.

Tell-tale signs of a mature design practice

- ☐ Stakeholders are able to relay the design team's strategy.
- ☐ It is clear how the design strategy relates to the wider organisational strategy.
- ☐ The design strategy is easily accessible to anyone who needs it.
- ☐ The design strategy evolves with other strategies.
- ☐ The design team references and advocates for the design strategy behind the scenes.

Potential barriers to scoring well

Partner teams do not have strategies to align with.

Tactics to consider:

- Position the design strategy as aligning with 'known' strategies.
- Ensure the strategy is flexible, so it can align in the future if things change.
- Don't be held back by others.

Partner team strategies have changed.

Tactics to consider:

- Work with partners to highlight the impacts of their changes.
- Understand why their strategy has changed and if the design strategy needs to change too.

Strategy: How & What

'Magical thinking' is the belief that someone's thoughts have an impact on the world, even if there is no clear connection between those beliefs and the changes that occur. In business, this can be observed when a team takes credit for a result without any clear connection. For example, a business has a strategy to reduce costs without any tactics to back it up. They feel that their strategy has worked when their costs reduce due to the price of raw materials dropping in the world market. Was the success due to a strategy without tactics or was it actually due to other actions out of their control? Should they continue with their current approach and hope for more 'magic' cost savings?

Although not unique to design strategies, an effective strategy covers:

- what is intended, and
- how it will achieve this.

For example, how will the strategic design goal of speeding up delivery time or improving the look and feel be achieved? Is it by reducing the approval process or creating a new design pattern? Without the 'how', even at a high level, the strategy is just a goal without a roadmap.

Having a clear set of 'how' tactics also enables the team to innovate by trying new approaches to see what does or doesn't work. If tactics don't work, the strategy can be updated with this information and shared with stakeholders to demonstrate progress, even if the 'what' and the 'why' of the strategy hasn't been achieved.

Identify what can or cannot be easily undone

Remember: One and two-way gates should be identified within the design strategy. These are the directions taken that are either:

- difficult to change (one-way gate), or
- easier to change later (two-way gates).

Being clear on this will help highlight the most upfront effort required to gather data before moving forward. An area of a strategy that's easy to change later will require less data to support it before proceeding.

Senior leaders will often want to understand this before agreeing to a strategy, as it helps them understand the risk level they are agreeing to.

Avoid a reliance on 'best practice'

Many design strategies struggle as they rely on 'this is best practice', but it's not a particularly convincing argument for investment.

Remember: 'Best practice' is widely used in design teams but often lacks substance. Apart from the largely ignored ISO, there is no 'Board of Design Best Practice'. Instead, differing opinions are common, and claiming a single best practice approach can mask a lack of deeper thinking.

Sakichi Toyoda's '5 whys' exercise is a useful tool for design teams to overcome their own bias and avoid terms such as 'best practice'.

Important: Getting to the 'why' of a design strategy is much more appealing to those who approve investments and are potentially forgoing/delaying other investment requests to prioritise design.

Set clear goals

It is far easier to get agreement and alignment in a strategy's 'how' and 'what' if the team adopts the business's existing language and measures. These, however, vary between organisations. Doing this helps people get up to speed more quickly and make connections to existing strategies.

Having goals such as Objectives and Key Results (OKRs) for team progress is a great tactic for communicating a design practice strategy and progress. If well structured, they can bridge the gap between setting up processes and achieving qualitative results. They are most powerful when used widely across the organisation and when the design strategy clearly relates to other organisational goals.

If OKRs are not used in an organisation, then other teams likely use something similar. It is worth spending time aligning the design strategy with the best and most accepted progress measures within the organisation.

Tell-tale signs of a mature design practice

- ☐ The design strategy has clear stepping stones to achieve goals.
- ☐ The design strategy has clear success measures/goals.
- ☐ The design strategy has process, tools, and/or staffing changes.
- ☐ The design strategy has flexibility to evolve as new information comes to hand.
- ☐ The design strategy references the predicted impact on the customer and business.

Potential barriers to scoring well

Too many unknowns.
Tactics to consider:

- Reduce unknowns with research or a research plan. Include more discovery work within the strategy.
- Provide different options once more information is known.

Lack of budget to fund the strategy.
Tactics to consider:

- Include ROI information in the strategy to assist with business cases for increased budget.
- Stagger the strategy to reduce budget impacts in a single period.
- Review what strategy elements are 'nice to haves' and 'must haves'.

Design is not expected to have a separate strategy.
Tactics to consider:

- Don't ask for permission. Create one and sell its value.

- Ensure the design strategy supports other existing strategies.
- If there is an existing strategy at a higher level, investigate a section for design within that broader strategy.

No time to create a design strategy.
Tactics to consider:

- Start small and expand over time.
- Review time-savings in design work; it is much easier to justify allocating time to the strategy if there will be a net saving in staffing.

Strategy: Investment

Most of us hold a personal investment portfolio in some way or another. We are constantly reviewing where we should invest our time and money.

- 'Should I be saving for a home, or should I invest in a holiday to improve my mental health?'
- 'Should I buy the raw ingredients to cook for myself, or should I get takeaway?'

We make hundreds of investment decisions every day because we're constantly considering the investment level required, the returns we'll gain, and how each decision impacts our overall life goals.

We also regularly make investment decisions in the workplace. Often, we think about how budgets impact personal goals rather than organisational goals. Personal and organisational goals can be aligned, but not always.

Important: Consider if we are recommending an investment based on what is best for our own career needs or the organisation's.

Size of team isn't everything

Design investment is where an organisation can go from talking about how important it is to *showing* how important it is. For some leaders, an investment mindset means adding more people to a team, for example, 'I have a team of two hundred designers, therefore I am invested in good design compared to a competitor who only has five designers'. Having more designers can be a great investment, but it can also be superficial.

Important: Hiring is just one part of a design investment budget. Tools, consultancy, training, and delivery investment are as, if not more, important than hiring.

Investments don't necessarily need to be in the design team itself. For example, investing in user-centred design training for product management can have more impact than hiring another designer. This is because it can make existing designers more efficient as they spend less time explaining their approach to product managers.

Investments are flexible

Investment budgets are sometimes thought of as 'how much money do I have to spend?'. Although budgets are often seen as finite, many executives state that there is often flexibility. Flexibility not only to reduce or reallocate budgets but to increase budgets. However, they really want to know the return on that investment. This is where a design team should always think about 'investments' rather than 'budgets'.

With the investment mindset, we highlight the time, effort, and costs in terms return on spend. If the return is high enough to offset the investment, coupled with the risk of that investment failing, then it's more likely to be understood and supported.

Important: Connecting an investment with the organisational strategy enables the team to show benefits beyond the products they currently design.

Good design strategy ROI not only shows the savings and revenue growth of what design teams are currently working on but also the impacts on other business areas. For example, investing in design thinking, service design, or experimental thinking can feed into ways of working across the organisation. Other returns include longer product shelf-life and improved SEO that often go hand-in-hand with greater design investments.

Prove your worth

To truly think with an investment mindset, it's important for design leaders to get more serious about proving their worth. Moving beyond the vague 'happier customers spend more' commercial arguments for design, it's important to have a mature strategy that leadership can get on board with. Design leaders now have more access to data that allows them to put a dollar figure on the revenue or usage uplifts their work has produced. AI tools have made it easier to access and combine data that would have previously taken specialist analysts a long time to compile. Embracing these new tools assists design leaders in moving beyond the vague arguments of the past to provide their strategies with more traction.

Own your budget

Many design teams feel they can't make the impact they're capable of as they're sometimes treated as a lower priority than other stakeholders. Design leaders are often at the mercy of other teams' budgets or are not seen as essential participants in budget discussions. Whilst finance and budget control are missing from many design leaders' expertise, it's something that design leaders can learn.

Important: Knowing how the budget game is played at your organisation is vital.

- What is flexible?
- What are the different processes?
- What are the budget timings?
- What evidence and scrutiny are required?

For example, managers oversee training budgets in some organisations, while in others they sit under central HR/P&C teams. At some organisations where there's a need to show savings each year, you can still grow the team and tools as long as those savings come through a reduced cost to serve customers. It's important to work out the nuances of both the official and observed budget process at your organisation.

Over investment is also a challenge

As frustrating as underinvestment is, design leaders should also be aware of the challenge of wasted spending, as it can lead to large cuts further down the line. Overspending without a clear return will lose senior stakeholder trust, making it harder to get more investment in the future.

Tell-tale signs of a mature design practice

- ☐ The design team has control over their budget.
- ☐ The design team budget is linked to those of the teams they work with.
- ☐ The budget includes ROI information.
- ☐ Investments include a mixture of training, tools, and staffing.
- ☐ Design investments go beyond the design team.

Potential barriers to scoring well

The design budget is unknown to the design team.
Tactics to consider:

- Find out who has this information and control in your business area.
- Be proactive with senior stakeholders about what information would help them.

The design budget is owned by projects or stakeholders rather than the design team.
Tactics to consider:

- Work closely with project/product managers and stakeholders to understand the current budget.
- Request a separate design practice budget for items that benefit all projects.

Budgets have already been assigned for the year/period.
Tactics to consider:

- Not all budgets work on a yearly cycle, so understand how out-of-cycle investments can be requested.
- Prepare for the next budget cycle to be ready ahead of time.

Lack of skills or awareness of the budgetary process within the design team.

Tactics to consider:

- Invest in training about budgeting.
- Increase the team's exposure to the budgeting process.
- Partner with finance to use their services.

Category: Craft

During the Industrial Revolution, manufacturing tasks were simplified so that individuals could become deep experts in a single area. This approach made sense in mass production given the lower levels of automation and education at the time. It meant quickly teaching someone to do a task well and then having them repeat it millions of times. Now, the landscape has changed with better automation and higher levels of education. That's why, for many years, businesses have been looking for T-shaped employees/teams, sometimes called 'unicorns'. The business logic is clear: Why have the overhead of multiple people/teams if we can have a single, flexible employee?

With the improvements in design tools, there is now a focus on visual design. It's a critical part of the design process because customers struggle if they don't immediately feel that something looks easy to use. However, with design tool improvements and use of repeatable design/code patterns, we should actually see less of a need for visual design skills, freeing up designers to develop other skills.

Important: The challenge for design leaders is that this can lead to doing a bit of everything badly or using methods without understanding the mindset behind them.

New tools, new opportunities

All teams have a set of processes, tools, and techniques that define their craft (hard skills). Stakeholders often assume this is all design teams focus on. In the extreme, a designer is only seen as someone with a good eye for style who knows how to use a design program like Figma. This narrow view can frustrate design teams, limiting their ability to demonstrate value or leading them to simply accept these misconceptions.

AI is making design more accessible to a wider audience while creating efficiencies for current designers, further reducing the effort required for visual design. Of course, views differ on how much designers should incorporate AI into their work and if 'design' can now be carried out by just anyone. Although AI lowers entry barriers into design, it doesn't mean just anyone will be a good designer. For current designers, AI can help shift focus toward design aspects that have sometimes been overlooked. This requires designers to learn even more methods and deepen their design knowledge.

Important: In the age of AI, the need for designers to focus on design skills is higher than ever; it's just that those skills are now broader and less focused on knowledge of one tool.

Stakeholders are often unaware of your design toolbox

Design craft is an important part of a design team, but it is a lot more complex than most non-designers realise. There isn't one method that fits all design challenges. Often called a 'designer's toolbox', a grab bag of different tools and outputs should be available to a design team in the same way a carpenter will not use the same approach to everything they make.

The three key sub-categories of craft focus are:

- Tools & Methods
- Delivery Link
- Types of Output

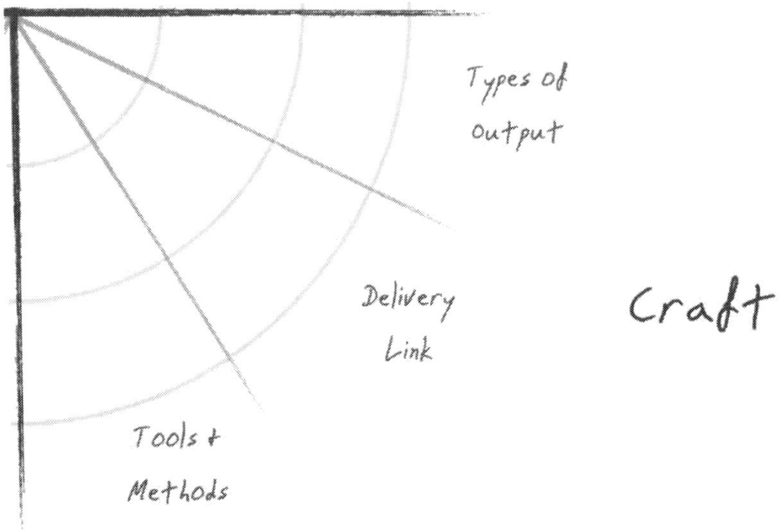

Craft: Tools & Methods

It seems like new digital tools hit the market every day that offer different features or open up methods to more designers and researchers. Many research and design methods have been around for decades. Methods that were unpopular in the past due to budgets or time constraints have become more appealing thanks to new tools and business models.

Improved tools and lower cost open up new opportunities

In the early 2000s, a usability testing lab may have cost US$150,000 to set up. A couple of years later, this dropped to US$50,000. Today, an in-person lab would be a fraction of that cost and include features only dreamt of back then. Similarly, in the last few years, remote user testing has not only dropped substantially in price but has also become much easier for researchers and participants to use. These changes have not only reduced the barrier to entry but also sped up the process, making it work much better with a rapid, iterative approach to product development. Not to say that there aren't still lingering beliefs that user research is an expensive and slow luxury, but designers now have fewer reasons not to test their designs early and often.

Digital design has seen the legacy print design tools swept away by new tools that enable reuse, development alignment, rapid prototyping, and better stakeholder engagement. Although most designers have moved to these tools, they often only use a fraction of their capability. Usually, this is because once a designer has learnt the tool, they're so focused on delivery that they don't explore the possibilities and new features as they're released.

Similarly, tools for A/B testing, analytics, personalisation, marketing, product management, and project management are now lower in cost and easier to use.

Important: Just as easy access to design tools doesn't make you a good designer, easy access to A/B testing tools doesn't make you good at experimentation.

For design practices to take advantage of these tools, designers need to broaden their knowledge of the fundamentals. Bringing a design mindset to other areas certainly has advantages but also requires designers to adopt the mindset and methods of these areas too. Otherwise, design practices will be no better than other practices that believe anyone can be a designer.

Time to learn

Improving the use of existing tools and methods or learning new ones takes time. Mature teams carve out the time to learn. Some teams struggle to account for upskilling. However, modern Agile processes that many design teams work in support upskilling. Developers will often allocate time in sprints for new tech training or spikes. Designers should do the same.

Important: In the race to learn new tools and methods, a design practice shouldn't dump all the old ones entirely.

Each tool and method have its pros and cons. Mature design practices will still use tools that provide value, such as wireframing and sometimes paper and pencil, even though some might think that high fidelity design tools have eliminated the need for these.

Understanding the tool and method needed for different challenges is a key part of a mature design practice.

Tell-tale signs of a mature design practice

- ☐ A review matrix exists to review new tools.
- ☐ Useful features of existing tools are quickly adopted.
- ☐ The team utilises a range of methods depending on the problem to be solved.
- ☐ Design delivery quality and speed continually improve.
- ☐ Time is allocated to reviewing and learning new tools and design methods.

Potential barriers to scoring well

Another team mandates tools or methods.

Tactics to consider:

- Collaborate with the other team to improve ways of working, including tools and methods used.
- Use case studies from other organisations and escalate with clear business cases.

Lack of time to investigate new tools and methods.

Tactics to consider:

- Work with stakeholders to carve out time using a business case that demonstrates the significance of future savings.

Lack of designer confidence to try new tools/methods.

Tactics to consider:

- Work new tools/methods into the designer's development plan.
- Find 'safe' projects for the designer to experiment on.

Craft: Delivery Link

As design tools have improved, development, marketing, content, and data tools have also improved. The barrier to entry has been reduced, so it's now easier for designers to understand their capabilities. The integration between business tools has generally improved, although it is far from perfect. AI is helping improve tool integration far beyond what has been possible in the past. But integration between tools is not enough; it also needs a strong collaboration framework between teams, safeguarding protocols, and shared processes.

Most designers care about what actually makes it into the customer's hands. A lot of frustration occurs when a designer hands off designs only to see something lesser delivered. A good understanding of delivery capabilities and working to improve the pipeline to the customer is a sign of a mature team.

Design systems

Design systems have recently had a lot of attention. UI design systematisation carries a lot of benefits to digital design practices and customers. The best design systems are those that are fully-aligned with the delivery tools, for example, code and testing with explicit links to the design system. This requires teams to work closely together not only to create shared repositories but also agree on processes and workflows.

Digital design doesn't stop when products go live

Content management is another area of delivery that can see substantial gains when linked to the product design process. Have you noticed user experiences that start out excellent but degrade over time as updates are made with less effort or skill? This reduces the commercial and customer impact of the original investment and shortens the product's shelf life. At a minimum, a mature design team will work closely with content management teams to ensure agreed standards are upheld.

This mindset applies to other downstream activities, whether it be an in-store pick up experience or customer support.

Important: Design practices should support the experience that the actual customer has rather than just hand over designs and hope someone else cares as much as they do.

Tell-tale signs of a mature design practice

- ☐ Design that the designer proposes and the final user experience delivered are similar, if not the same.
- ☐ Design and delivery teams share initiatives to improve speed and/or quality.
- ☐ Design solutions are within time and budget constraints.
- ☐ User experiences don't degrade over time in-market.
- ☐ Suggested designs account for in-market management.

Potential barriers to scoring well

Delivery teams are unwilling to change their ways of working.
Tactics to consider:

- Work with the teams and their leadership to agree on goals that can only be reached through change.
- Share case studies of others' success.
- Celebrate when ways of working improve.

Lack of time or connection with the delivery teams.
Tactics to consider:

- Build good working relationships/partnerships with these teams over time.
- Schedule regular catch-ups to get to know the teams better.

Legacy systems restrict delivery links.
Tactics to consider:

- Look for opportunities such as upcoming re-platforming/redesign projects to include work that better links design and delivery.

Craft: Types of Output

Craftspeople will constantly change their output. For example, potters may start their career making one-off pieces then move to producing more of a successful design and eventually create a mould to replicate that design en masse. Each step is a refinement of what they are producing. However, their current ways of working can sometimes hold them back from their goals. In this example, a potter may also continue to create one-off pieces, but to increase their income, they may also produce social media and other marketing content, imparting more value to their one-off pottery items. Either way, maturing their practices will lead to more diverse and refined outputs.

Some design teams' output has been fairly stagnant for decades. Those outputs will often take the form of a handoff to other teams in the organisation. This has been perpetuated by the attitude of the teams and organisations they work in.

Work with partners to improve output

Examining handoffs to others is often given little consideration. 'Why fix what isn't broken, hey?!' This allows inefficiency to go unchecked.

Important: Regular check-ins with partners to identify friction can strengthen bonds between teams and frees up time to focus on value-added activities.
An example is that there's less need to hand high-fidelity mock-ups to developers when a strong digital design system is in place, as they already know the modules in the design system.

Mature design practices will be constantly looking for efficiencies in how they work with others. This gives them case studies they can then share with stakeholders, which in turn leads to them being included in future discussions when there's a need to refine ways of working.

Find gaps and fill them to add value

Evaluating output also enables a team to expand its impact by identifying gaps and filling them to add additional value to the organisation. For example, service design will often grow out of a UX team that will take on new types of output, like service blueprints, that help them and other teams understand the systems and processes.

Important: Such moves by design practices have often helped organisations better understand the benefits of investing in design beyond how things look.

Tell-tale signs of a mature design practice

- ☐ Partner teams are consulted about how processes can be improved.
- ☐ Design briefs don't dictate the methods and outputs.
- ☐ Designs are well understood by delivery teams.
- ☐ Partner teams and stakeholders do not expect a design team to just create high-fidelity designs.
- ☐ Output standards do not vary greatly between designers across the practice.

Potential barriers to scoring well

Another team sets the output.
Tactics to consider:

- Work with that team to understand their needs and the reasons for their demands.
- Create a business case for change, ideally with that team.

A change in output would have knock-on impacts on other teams.
Tactics to consider:

- Speak to all teams that could be impacted to understand the pros and cons of your changes.
- Create a business case for change to understand if the cost of change outweighs the cost of not changing.

Lack of knowledge about potential output improvements.
Tactics to consider:

- Share case studies from other organisations.
- Run low-risk trials to demonstrate the changes and measure impacts.

Category: Reliability

It's easy to forget that business leaders face the same human challenges as everyone else. As part of the human condition, they seek predictable and reliable outcomes. If they invest in X, then Y is the likely outcome. Unfortunately, design leaders sometimes send the message that 'design is messy', scaring business leaders. That's why designers should consider the psychological perspective behind business leaders and, in conversation, frame design as a predictable activity.

Any team can get lucky and have great results, but making results reliable in the future shows real value and ensures that success isn't just built on an individual designer.

Look to the future

A mature design practice will not only focus on the initiatives they are working on today but also on how future initiatives are set up for success.

Important: Many senior leaders state that reliability is critical in their decision making, so having a reliability focus within the practice can also ensure that the design team is included in higher-level discussions.

Higher confidence, higher investment

Reliability reduces risks and increases confidence in design practice investment. It can be seen in business throughout the world that an outcome's reduced risk increases investment. This doesn't mean that a focus on reliability leads to a lack of focus on innovation. In fact, many successful design practices that carry high levels of reliability are trusted to take on the challenge of innovation.

The three key sub-categories of reliability focus are:

- Workload Management
- Success Measures
- Improvement Plans

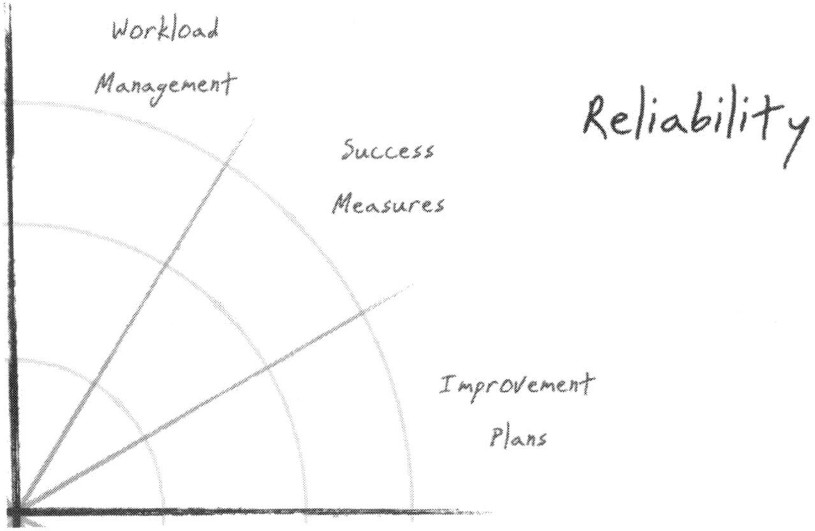

Reliability: Workload Management

Design leaders often say that managing a design practice feels like spinning plates. That sounds scary! It certainly is for an amateur, but a professional plate-spinner (yes, they do exist) won't break a sweat when spinning away on stage. A professional knows which plates need attention, the different techniques to keep them spinning just right, and can easily add more plates to the mix. They've built up this ability through practice and by investing in reliable tools.

A challenge for all teams, especially those embedded in Agile delivery squads, is that work can be unevenly distributed amongst a team or there is fluctuating demand for specific skills. Some designers will be spinning too many plates to keep track of, while others are looking for more plates. So, how do you balance these demands? This is much easier to address with a central, highly experienced practice. The downside of centralised design teams is that they can struggle to collaborate meaningfully with delivery squads. This is why some organisations have flip-flopped between the central and embedded team models.

Knowing if things are on track

Tracking workload and progress is an issue many design leaders raise. Designers aren't always focused or motivated to update information about their progress. There can often be challenges with less mature designers being unable to break down their work into trackable tasks. A mature design practice will have support for designers to break down their tasks and report on progress.

Using workflow tools enables a leader to assign or reassign resources to new initiatives. They also enable the practice to easily share what they are working on and their capacity. A great head start is when the organisation has high-quality workflow tools or flexibility in what teams use. However, some organisations have strict policies on workflow tools. This is understandable, as interoperability between tools is often lacking and stakeholders don't want to swap between many tools to track what different teams are working on.

Unfortunately, where a single workflow tool is mandated for all teams, it often won't meet a design practice's needs. In these situations, the design team needs to adopt the tool and also work on a business case for improving it.

Plan for improving the design practice

It can be a challenge to improve a practice when resource tracking is based solely on project delivery. Ideally, there'll be an understanding that time is spent improving skills and the overall practice. If this understanding doesn't exist in the organisation, skill and practice improvements may be able to be attached to related initiatives. For example, a re-platforming project may take on the additional tasks of training team members and setting up a design system that other teams also use. Or a CMS project may take on the additional tasks of creating a service blueprint that other initiatives also use in the future.

A key challenge in these situations is that initiative sponsors may not understand or protect this scope of work. This is why mature practices have a greater level of control over the balance of time spent improving the practice and delivering initiatives.

Establish an overflow mechanism

It's important to reduce risks when new projects are quickly spun up and exceed the existing design practice's capacity or capabilities.

Important: A mature design practice will have agreements in place with external resource providers.

This reduces the chances of inappropriate hiring/outsourcing and increases alignment between teams.

Tell-tale signs of a mature design practice

- ☐ It's easy for anyone to know what the design team is working on.
- ☐ Time is allocated for training, not just project/initiative work.
- ☐ The team can rapidly scale up when required.
- ☐ Team members can move between initiatives without issue.
- ☐ Team members and stakeholders do not feel any designer is over or under-utilised.

Potential barriers to scoring well

Do not have a reliable agency available at short notice.

Tactics to consider:

- Work with recruitment consultants to build a list of reliable freelance talent.
- Spend time analysing potential agencies and their talent before engaging them.

The budget does not allow for flexibility within the team.

Tactics to consider:

- If a new initiative is spun up that requires extra resources, feed this into the business case to ensure the budget.
- Work with stakeholders to understand what current work can be deprioritised to enable the team to focus on new projects.

The organisational/project structure does not allow for flexibility in workloads across the team.

Tactics to consider:

- Work with stakeholders to understand how work can be allocated to teams with the right level of design support.
- Align design timelines with the level of design support in that team.

Reliability: Success Measures

Someone putting a bet on a horse is about looking at past performance and only risking big money when indications say that the horse will deliver again. How do they know? They use success measures.

Success measures for a design practice provide better confidence in repeatable successes and self-correction when needed. Success measures can vary between organisations. Whatever the measure, it should be accounted for early on, be comparable to past performance, be shared, and be trusted. The latter is vital, as untrusted measures are worse than no measures.

If they don't exist, create them

In situations where product teams don't have clear success measures (which happens more than you'd think), an immature design practice will be frustrated but take no action. This is when the designer or design leader should take the initiative to set success measures. These could be organisational (for example, throughput) or commercial (for example, revenue uplift).

Important: Success measures should align with the organisation's strategy.

At a practice level, success measures enable a design leader to understand where things are going well and where more attention is needed. There can be many reasons for success not being achieved, and it may have nothing to do with the design practice. Having the measures in place and regularly reporting on them enables the design practice to get out ahead of any issues before they are repeated.

Reporting on measures shows seriousness

Reporting on success measures addresses a common complaint about design practices: that they are often unserious about commercial matters or struggle to quantify their benefits. These complaints are far less often laid at the feet of development teams, as Agile development teams have given them success measures (bugs and throughput/velocity) and a reporting framework.

Important: Although design practices will have different measures, they should be kept simple and consistent like developers do.

Tell-tale signs of a mature design practice

- ☐ Customer/user experience measures are in place and reported on regularly.
- ☐ Team efficiency and quality measures are in place and reported on regularly.
- ☐ The team has clear targets/goals.
- ☐ Stakeholders and partner teams see the impact of changes in the design practice's ways of working.
- ☐ Individual team members understand where they are excelling or underachieving.

Potential barriers to scoring well

Lack of access to data for success measures.

Tactics to consider:

- Create a business case to set up your own success reporting.
- Escalate to gain access to existing data.
- Start with small/basic measures and improve over time.

The team has moved on to the next initiative without reflecting on previous delivery success.

Tactics to consider:

- Work with the team to establish a Definition of Done (DoD) that includes a review of success.
- Collaborate with stakeholders to change ways of working.
- Include time into any initiative planning for post-release reviews.

Reliability: Improvement Plans

There is always room for improvement for individuals and practices. Being aware of those areas and being honest with stakeholders is important for long-term success. It's better to identify areas of improvement yourself and create a plan to tackle them rather than letting someone else identify them and control the narrative. Being open also makes it easier to request additional investment in the future, if required.

If the organisation has existing individual development plans, consider building on them as they may be generalised to fit any employee or linked directly to a benefit scheme.

Important: Use improvement plans consistently and regularly check in to keep them updated. This will ensure that team individuals are building skills and experience in the areas most beneficial to them and the team. It also reassures others in the team that continuous skills improvement is expected.

Don't try to improve everything all at once

Implementing focused and manageable improvement plans is another important factor.

Remember: Trying to improve everything all at once is likely to lead to clashes and poor progress. A strategy that breaks improvements into different time horizons will not only assist the team but also reduce noise from stakeholders who are curious about when improvements will be seen.

Coupling improvement plans with success measures enables progress to be tracked and shared.

Tell-tale signs of a mature design practice

- ☐ Individual team member improvement plans are in place and aligned with the design strategy.
- ☐ Team improvement plans are included in the design strategy.
- ☐ Team improvement plans have several time horizons.
- ☐ Stakeholders are aware of the areas of improvement that the practice is focusing on.
- ☐ Improvement plans are linked to the organisation's strategy.

Potential barriers to scoring well

Difficult to carve out time.

Tactics to consider:

- Only use 80% of a designer's time for initiatives, keeping the rest clear for self-development and practice work.
- Increase resources or look for efficiencies in other work.

Design is distributed, so it's difficult to implement a single improvement plan.

Tactics to consider:

- Look for common issues across teams that, if resolved, will deliver clear value.
- Have a practice improvement plan that calls out different opportunities within each team.

The interests of the team members' improvement and the needs of the organisation aren't aligned.

Tactics to consider:

- When hiring new designers, consider what organisational needs aren't currently being met.
- Work with the team to highlight the organisational needs, so they have better awareness of future opportunities.

Chapter 5: Worked examples of the design practice conversation framework

To help build an understanding of how design leaders can apply the conversation frameworks, we'll run through three example scenarios of different practices.

These include:

- Brief background about the situation.
- How the design practice scored against the framework.
- Challenges the design practice faced.
- Areas they selected to tackle first.
- The conversations they had using the framework.

Scenario 1: Design strategy was overlooked

Background

- Newly formed product design team.
- SAS company that is rapidly growing in a niche sector.
- All designers report to the same manager.
- A new product project has been underway for a few months.

Problem to solve

A design strategy wasn't something the organisation was familiar with or how it differed from a product strategy.

Current success measures

- Belief in design team impact: Mid.
- Customer satisfaction: Low.

Forum for discussions

- 1-to-1s with stakeholders.
- Team day.
- Leadership forum.
- Monthly update newsletter.

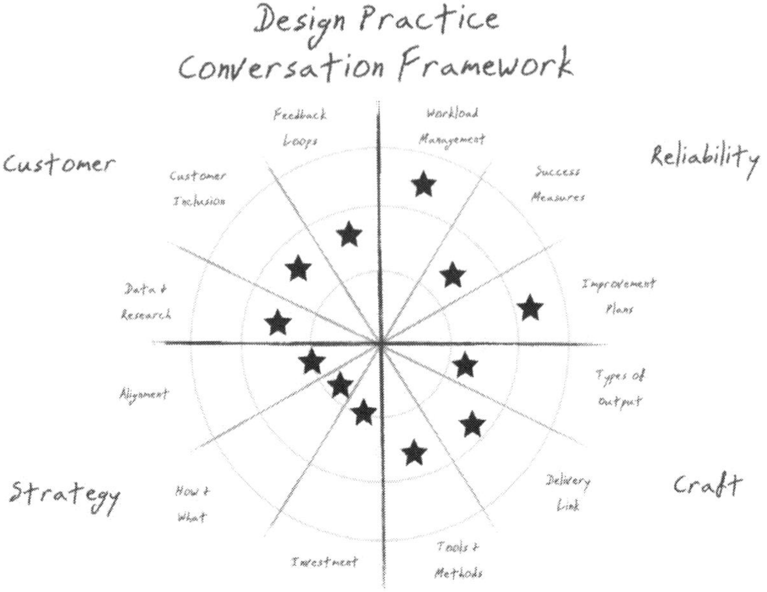

Approach taken

Stakeholder engagement and alignment

- The design leader met 1-to-1 with their line manager and the heads of IT and product.
- They used the design practice conversation framework to identify and discuss strengths and weaknesses.
- They explained what a design strategy entails, its value, and how it aligns with company, engineering, and product strategies.
- They also explained what's necessary for quality design success measures, as this area was less familiar to the head of engineering.

Team reflection

- The design practice conversation framework guided a team day focused on reflecting where the design practice was strong or weak.
- The team agreed that success measures and strategy were important and started gathering information and ideas to focus on.
- They were also interested in other areas of opportunity but understood these would be addressed in the future.

Action plan

- The design leader turned the insights into a 3-month action plan.
- They presented this at a business leadership forum.
- They were clear to highlight an initial focus on three key areas, with plans to build on them over time.
- There were a few questions about the areas not included, and there was general agreement that not everything could be addressed at once.

Keeping stakeholders informed

- In the monthly design newsletter, the design leader shared the plan to address the three key areas and committed to providing progress updates in future editions.

Focus area for coming quarter

- Reliability: Success Measures
- Strategy: How & What and Alignment

Scenario 2: Continuously revisiting design conversations

Background

- An established team within an ecommerce company.
- High financial investment.
- All designers report to the same manager.
- Design lead has used the design practice conversation framework in the past to agree in which areas of the practice to focus uplift efforts on (Reliability: Improvement Plans, and Craft: Delivery Link).

Problem to solve

The design leader wanted to build on the momentum of making initial improvements to the design practice. Progress had been made but they felt there was still lots more to do.

Current success measures

- Belief in design team impact: Mid-Low.
- Customer satisfaction: Unknown.
- Team throughput: High.

Forum for discussions

- 1-to-1s with stakeholders.
- Team day.

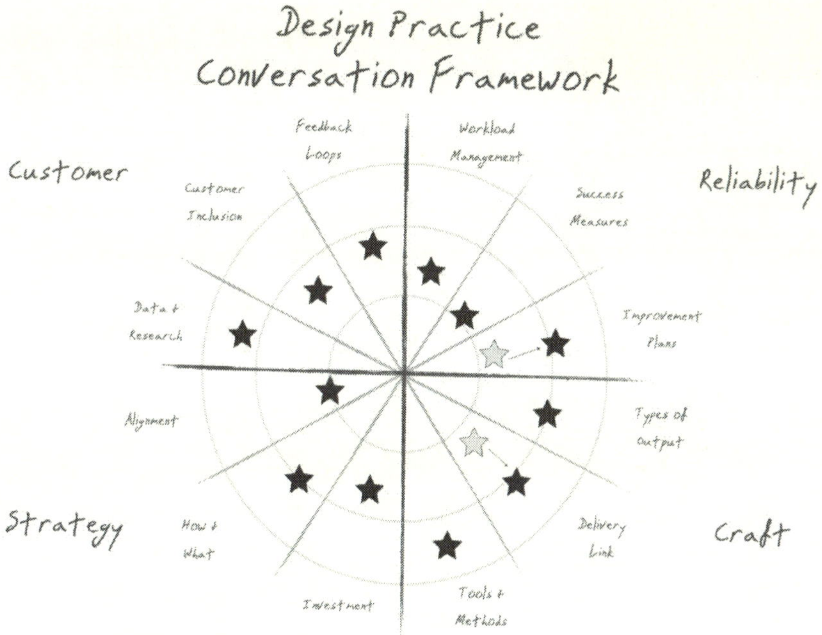

Approach taken

Design practice health check

- During a team day, the design leader ran a session to explore design practice strengths and weaknesses.
- They used the design practice conversation framework to guide the discussion.
- This brought up a number of areas of pride and concern.
- It enabled the team to reflect on the improvements they had recently made to the design practice, which was included in the conversation framework.
- One issue was that new features were often delivered without checking if they were successful.

- Another area was that the design strategy from the previous year hadn't gained traction, as product and marketing teams weren't aligned with it.

Aligning stakeholders

- With this information, the design leader set up 1-to-1s with stakeholders who could help measure and report on the success of their work.
- They also met with the product and marketing leaders to revisit the design strategy produced the previous year.
- Over several sessions, the strategy was updated to better align with the product and marketing teams' evolving plans.
- The product leader invited the design leader to a monthly session with their team to present on design practice health, using the design practice conversation framework to facilitate discussions. This included information about recent improvements the team had made.
- This led to a valuable discussion about workload management, as the product team was also struggling to assess this across Agile teams.

Focus area for coming quarter

- Reliability: Success Measures
- Strategy: Alignment

Scenario 3: Discussing fragmented practice improvements

Background

- Several groups of designers reporting into different departments (fragmented).
- Highly capable designers.
- When new initiatives launch, project managers often bring in agency designers.
- There is an Agile guild for designers to meet up.

Problem to solve

A line manager asked one of their design leads to address concerns about the low quality of design across the organisation.

Current success measures

- Belief in design team impact: Low.
- Customer satisfaction: Unknown.
- Team throughput: High.

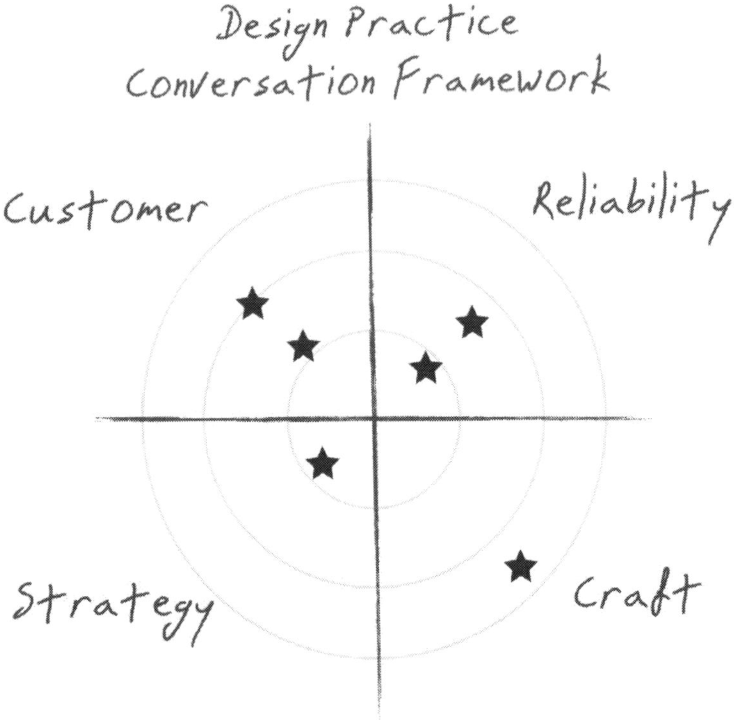

Approach taken

Understanding the current state

- The design leader worked with design guild colleagues to gather information about current practices across the fragmented design practice, identifying what was going well and what wasn't.
- They used the design practice conversation framework for these discussions, as it was a neutral structure that didn't show bias to any particular design team. The other design leads were also keen to take part in the discussions.

Differentiating strengths and weaknesses

- Given the different ways the design teams operated, there was a need to differentiate their strengths and weaknesses.
- They chose to use the high-level version of the design practice conversation framework, as a quick turnaround was needed, with the intention to thoroughly revisit the detail in the future.

Presenting findings and next steps

- The design leader and their colleagues presented the overall design practice's strengths and weaknesses to senior leaders.
- This helped the senior leadership team understand what support the design practice required to improve.
- They agreed the initial focus should be on improving reliability, followed by a heavy focus on creating a consistent approach to designing with customers in mind.

Forum for discussions

- Design guild meetings.
- Senior leadership meetings.

Focus area for coming quarter:

- Reliability, particularly in the area of the fragmented design practice that was scoring very low here.

Chapter 6: Applying the Individual Designer Conversation Framework

The framework supports honest reflection on strengths and growth areas, useful for both professional development and hiring discussions.

This framework enables free-flowing conversations for both current and potential employees. It's hard to guess what a designer is thinking and the area they feel is important to focus on. The framework helps reduce guessing, so there can be open conversations about the designer's current skills and where they want to grow next.

Important: It also ensures those conversations are consistent over time and with other designers in the team.

The framework has four categories:

- Technical Mastery
- Solution-focused
- Collaborative
- Organised

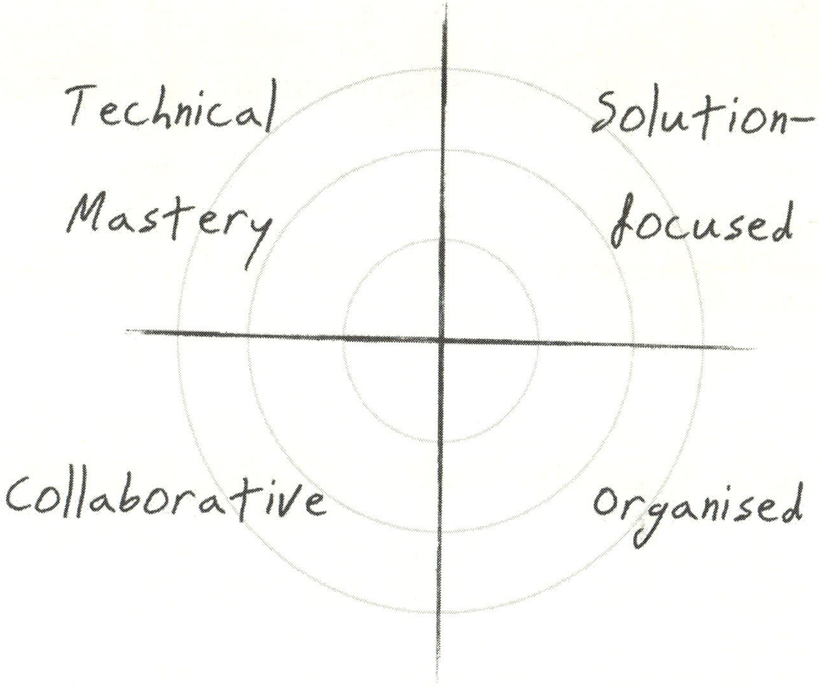
Technical Mastery
Solution-focused
Collaborative
Organised

Category: Technical Mastery

Masters of any discipline have similar characteristics, whether it be Bruce Lee, a master of several martial arts, or Pablo Picasso, a master of modern art. They don't limit their knowledge to their original area of focus; they are constantly learning, and they adjust their approach depending on the situation. They are technical masters of their area not because they know how to do one thing well but because they have such technical mastery that they can apply their skills flexibly.

Technical mastery covers the tools needed for the role. This isn't just limited to design applications; it also includes the different methods that can be used (tool application). Often called the designer's toolbox, a solid understanding of what method to use in different situations is key to being a productive designer. Tools have and will continue to evolve, so focusing only on what is needed today isn't necessarily a recipe for future success. A mature designer can usually pick up a new tool and be effective with it in a matter of days.

A designer should have a level of technical mastery that matches their role seniority. The range of tools they have mastered will usually increase over time, as well as their ability to quickly pick up new tools and techniques. So, a junior designer may have a lower level than a senior in the designer conversation framework, which isn't a bad thing as it's simply aligning with the expectations of the role seniority.

Technical Mastery is broken down into:

- Understanding of Other Disciplines
- Continuous Learning
- Right Tools & Methods

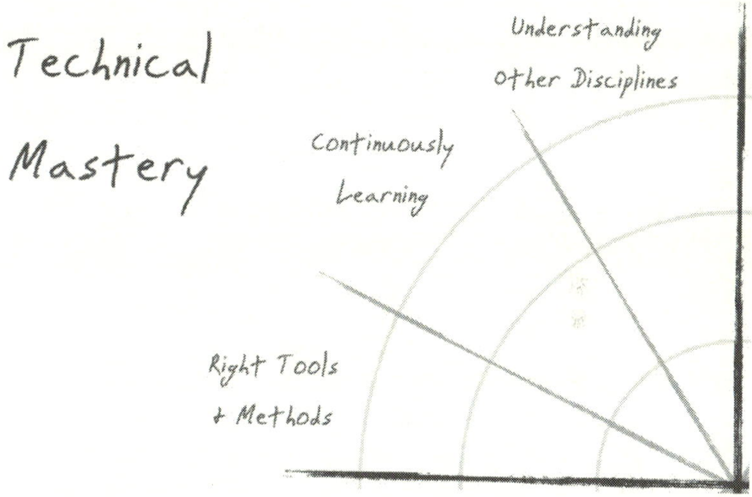

Technical Mastery: Understanding of Other Disciplines

Why is an understanding of other disciplines important for technical mastery?

Design is not a solo activity. A team member works with other designers as well as developers, marketers, business managers, analysts, and others. Having an understanding of these other areas helps the designer perform and collaborate better. This doesn't mean they need to be an expert in these other disciplines. If they feel like they are, it can lead to clashes.

Don't limit your T

T-shaped designers (those with a deep knowledge of an area and broad knowledge of other areas) have historically been limited in their skill sets. Often, the expectation is to have broad knowledge but in limited fields, such as other design disciplines, research, or programming. In reality, there are many areas where it's useful for designers to have knowledge. Analytics, experimentation, product management, marketing, copywriting, business requirements gathering, architecture, and business strategy are all areas in which a designer can leverage their existing skills.

Asks questions

A mature designer will be keen to better understand the challenges their colleagues face. They'll ask questions to learn more, have regular catch-ups with colleagues, attend broader training, and seek out opportunities to practice their new understanding.

Important: Having an understanding of the other disciplines a designer works with improves their output quality. They'll be able to ask better questions, build better relationships, and produce solutions that take advantage of more opportunities.

Tell-tale signs of a mature designer:

- ☐ Regular catch-ups with other disciplines.
- ☐ The designer references opportunities from other disciplines by connecting the dots between their work and others'.
- ☐ The range of work the designer takes on is constantly broadening.
- ☐ The designer receives positive feedback from colleagues in other disciplines.
- ☐ The designer seeks training in other areas.

Potential barriers to scoring well

Lacks familiarity with basic concepts used by other disciplines.

Tactics to consider:

- Task the designer with leading the learning for the whole practice of selected areas from other disciplines.
- Partner the designer with a specialist in this area and/or work on initiatives that expose them to these concepts.
- Support them to create a learning plan.

Lacks openness by other disciplines.

Tactics to consider:

- Work with senior leaders to open up other disciplines for sharing.
- The designer arranges catch-ups with potential partners in other disciplines to better understand their work.
- Use training and AI to gain an understanding.

Lacks interest in learning about other disciplines.

Tactics to consider:

- Understand what motivates the designer and align that with relevant disciplines.
- Provide the designer with a clear goal that requires a broader understanding.
- Invite other disciplines to present to the team.

Technical Mastery: Continuously Learning

The term, 'Half-life of facts', is the concept that all knowledge eventually becomes outdated. In all professions, from medicine to engineering, updated information replaces what is learned in training or early career. In the same way a doctor treats a cancer patient differently than they did ten years ago, a designer creates a mobile interface differently than they did ten years ago. It's important for designers to stay current, not only with changes in tools, design patterns, and technology but also with changes in customer preferences. What customers like, dislike, expect, or disregard is constantly evolving as their needs and experiences change.

There are always new skills and approaches to learn. The range and capability of technical tools appear to be increasing, particularly with all the possibilities of AI to aid design and research. A high-performing team member will be actively aware of the need to keep reviewing their knowledge and skills.

Learning can be individual but is even more impactful together

Designers often partake in sociocultural learning (learning with others) and observational learning (learning from others) to broaden their knowledge.

Many designers encounter similar challenges, so learning from other designers' approaches and thought processes can provide a deeper understanding compared to solo theoretical learning. Individuals dedicated to continuous learning often seek out others to learn with and from. Vygotsky's concepts in this area offer useful insight into the power of sociocultural learning.

Important: The benefit for organisations of shared learning is that improved knowledge and capabilities are shared across employees, thus reducing the risk of capabilities walking out of the door with the employee one day.

Tell-tale signs of a mature designer:

- ☐ Shares information they've learned with others.
- ☐ Keeps across new tools and features that could improve their workflow.
- ☐ Has a proactive development plan that stretches them.
- ☐ Proactively requests development opportunities.
- ☐ Will actively listen to others who are sharing knowledge and skills.

Potential barriers to scoring well

No time to learn new skills.

Tactic to consider:

- Work with the designer to manage workload, deadlines, and efficiencies. This highlights if they lack time and need support or are choosing not to prioritise learning.

The designer's tools and methods are restricted.

Tactic to consider:

- Identify opportunities to trial new technologies at little or no cost.
- Break down opportunities to minimise risk and cost rather than a big bang approach.

No budget for training.

Tactics to consider:

- Use mentors.
- Find 'safe' projects to enable learning through doing.
- Use free online/AI learning sources to support them.
- Secure training budgets.

Attitude of superiority reduces openness to new learning.

Tactics to consider:

- Focus conversations on what 'they' can do to improve outputs/ways of working.
- Ensure conversations avoid what 'others' can do, as this detracts from introspection.
- The best designers will be looking to constantly build on their existing knowledge/skills.

Technical Mastery: Right Tools & Methods

Every good designer will have a toolbox of different tools and methods that they use depending on the type of design problem to solve, the audience, or a project's phase. Having a wide range of tools and methods is important, and so is knowing the best time to use each one.

Risk of knowledge stagnation

Important: Experienced designers can become stagnant in their knowledge if they only focus on familiar tools and discrete aspects of design.

The ability to continue to learn new ways of working and being aware of new possibilities with existing tools is vital for experienced designers.

Avoid jumping onto every new thing

Less experienced team members can jump to new tools quickly as they are less invested in their current ways of working. Although this can be a great asset, it can also lead to issues if the new tools don't support the design process's core needs. Encouraging a designer to actively look for new tools and methods to assist their work should go hand-in-hand with them having an understanding of the core issues they are trying to address.

Understand the difference between executing poorly and method shortcomings

The internet is full of articles about method X or Y being worthless. These often give examples of a method being executed poorly and thus having a poor result. It is absolutely possible and frequent that the wrong method is chosen. It is also true that all methods have their shortcomings. However, any method executed badly will have low quality outputs. It is far too easy for designers to see a poorly executed method and dismiss the method itself. A mature designer will not only know the most effective way of executing methods but also understand the strengths and weaknesses of each.

Mixing methods/tools

During the life of a design initiative, a designer will use a collection of different tools and methods. A mature designer will know what methods will work well together (or not) for the initiative. Using a poor choice of tools/methods will lead to gaps or overlap and likely confuse stakeholders. This is often seen when a designer uses the same set of methods for every initiative, no matter the problem to be solved.

Tell-tale signs of a mature designer:

- ☐ Lists the key challenges of the particular initiative they are working on to form their design plan.
- ☐ Is able to clearly articulate why they have or haven't chosen a particular tool/method.
- ☐ Shares information about new tools/features/approaches with others.
- ☐ Reads/listens widely to understand the approaches of other designers.
- ☐ Avoids common pitfalls of the tools/methods they use.

Potential barriers to scoring well

Limited understanding of tools/methods.

Tactics to consider:

- Create a development plan with the designer with goals to improve their knowledge. This should include formal learning, practical application projects, and partnering with mature designers.

Limited practical experience using key tools/methods.

Tactics to consider:

- Identify projects that enable them to apply the key tools/methods that they have a theoretical understanding of.

Only comfortable with a small range of tools/methods.

Tactics to consider:

- Identify 'safe' projects that the designer can learn and try out new methods/tools on.
- Partner the designer with more experienced designers who they can shadow.

Designer has extreme views about particular methods/tools.

Tactics to consider:

- Set up mentoring with more mature designers/leaders who have a more nuanced understanding of tools/methods.
- Challenge the designer to identify what would change their mind to understand if their views are well-founded or dogmatic.

Unwilling to try something new.

Tactics to consider:

- Create a development plan with clear goals for identifying and trying new methods.
- Gain an understanding of the unwillingness to see if this is based on mindset, workload, etc.

Category: Solution Focused

Designers are great at uncovering issues and opportunities. Moving from admiring a problem to having effective solutions is the hallmark of a great designer. 'Effective' solutions should be the key focus.

Leaders highlight the importance they put on designers not just identifying issues but coming up with solutions and options.

Important: Do not lose sight of the problem when focusing on solutions.

Being solution-focused means going beyond just what's been briefed. Fully understanding the problem at hand ensures that the design solution meets business and customer needs, especially when these are not immediately obvious.

Being solution focused is broken down into:

- Insight Hungry
- Customer Focused
- Commercially Minded

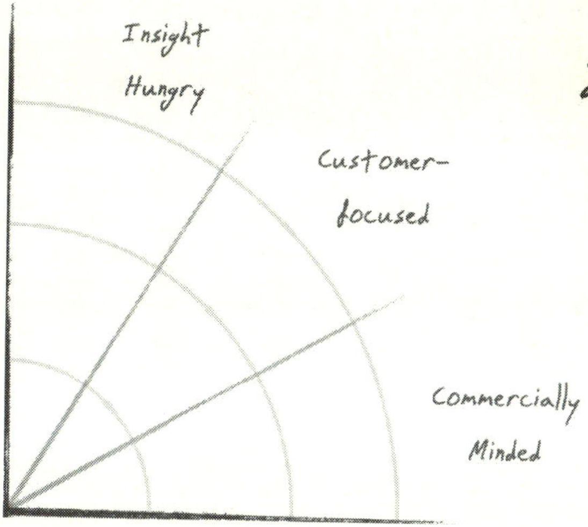

Solution-focused

Solution Focused: Insight Hungry

To create effective solutions, a designer needs to be hungry for understanding of customers, technical possibilities, commerciality, legal considerations, historical data, and so on. Having this hunger leads to well-rounded solutions and fewer late-stage challenges.

There are many ways to be insight hungry, but the most obvious one is to regularly ask for feedback and test any actions taken from it.

Expectations are always rising

When a designer lacks curiosity about their work and stakeholders, their work can plateau and deteriorate over time. Work standards and expectations gradually increase, and if the individual is not aware of this or is ahead of the game, their work's perceived value may reduce.

Be experimental

Having an experimental mindset is particularly useful for a designer. An effective designer maintains this mindset by gathering feedback on each iteration of a new idea then refining and gathering feedback on the next iteration.

Important: Being insight hungry will also enable designers to understand when a concept is not worth pursuing further.
This reduces time and money spent on low-value concepts and frees up the designer to think of new ideas that better align with organisational goals.

Tell-tale signs of a mature designer:

- ☐ Justifies design decisions based on customer insights.
- ☐ Justifies design decisions based on stakeholder feedback.
- ☐ Will plan for multiple iterations of designs, so they can act on insights.
- ☐ Will test several concepts and kill poor-performing ones.
- ☐ Runs workshops with stakeholders to gain early feedback in projects.

Potential barriers to scoring well

Lacks access to customer insights.
Tactics to consider:

- Work with the designer to identify how they might gather customer insights and who in the organisation can assist them.
- Help them understand that insights can be gathered rapidly at low cost if required.

Lacks access to stakeholder feedback.
Tactics to consider:

- Look for forums such as showcases or retros to gather feedback.
- If these don't exist, the designer can set up a time with stakeholders.

Solution Focused: Commercially Minded

An area that everyone can align on is that they want their organisations to succeed. Using the language of commercial success can frame design ideas in a positive way as stakeholders can easily align them with other work to deliver the organisation's goals.

A good designer will be able to frame their ideas in a commercial landscape. An important part of that is including the 'so what' of their ideas. What bottom-line impact will the design have or what could be the negative impacts of not acting?

Avoiding low-value ideas

Being commercially minded will enable the designer to better understand when to avoid lower-value ideas.

Important: Choosing which battles to fight improves design effectiveness, as over time, stakeholders will become aware that the designer highlighting an issue or opportunity means it's likely commercially important.

Commercials are not one-size-fits-all

Commercials vary between organisations and industries, so it is important for individuals to be across the organisational strategy and commercial drivers. Designers are unlikely to get cut through if they highlight the wrong commercial factors.

Important: Financial commercial goals often go beyond just revenue as they can be profit or future profit-focused, requiring a focus on costs.

It is important to remember that most organisations' commercial goals are not just financial. Many will have social, employment, and regulatory goals that form part of their commercial strategies.

Tell-tale signs of a mature designer:

- ☐ Regularly presents designs with reference to how they align with commercial goals.
- ☐ Will look to understand the cost to create and maintain solutions they propose.
- ☐ Will kill their ideas that don't have a commercial impact.
- ☐ Keeps up with any changes in the organisation's strategy and focus, as well as those of competitors.
- ☐ Will consistently report on their work's success or failure in a commercial light.

Potential barriers to scoring well

Commercial objectives are not available to them.

Tactics to consider:

- If commercial objectives are truly not available, this should be escalated.
- If they are just hard to get hold of, point the designer in the right direction.

Commercial strategy is very high-level and hard to align with.

Tactics to consider:

- Assist the designer to bridge the gap.
- Request that a business leader run through the strategy with the designer/team.

Solution Focused: Customer Focused

When individuals are truly customer focused, they understand that many teams impact the customer, not just their own. Being aware of the customer's end-to-end experience and the impact that different teams have on it helps a designer create more impactful designs.

In a designer's own decisions, they should have the customer in mind. This requires a good understanding of the target customer groups and building empathy for them. Being customer-focused increases the likelihood of ideas meeting customer needs and reduces the distraction of low-impact ideas.

Research, research, research

Some designers carry out their own customer research and some rely on the research of others. If research doesn't exist, a mature designer will know how to get it commissioned. They will also be aware of the different types of customer insights available at different stages of the design process. Customer-focused designers will be strong supporters of using customer insights and data throughout their design process and be able to justify its need when challenged.

Keeping the customer front of mind

Unfortunately, designers and teams can sometimes lose focus on the customer as a project progresses and important decisions will be made throughout the project lifecycle.

Important: The designer should take responsibility for maintaining customer focus.

Tell-tale signs of a mature designer:

- ☐ They consistently reference the customer and their needs throughout a project.
- ☐ They consistently reference customer research rather than personal opinion.
- ☐ If customer research/insights are not readily available, they will go the extra mile to obtain them.
- ☐ Stakeholders look to them as a source of information about the customer.
- ☐ They consistently draw a clear line between the customer's needs and what they've designed.

Potential barriers to scoring well

Lacks access to customer insights.

Tactics to consider:

- Escalate to gain access to existing insights.
- Provide resources to the designer so that they can gather customer insights.

Project teams or stakeholders are disinterested in customer needs.

Tactics to consider:

- Encourage designers to bring the customers to life using tools such as customer interviews.
- Work with the designer and business to link customer needs to commercial success.

Differing views of who the target customers are.
Tactics to consider:

- Encourage designers to be explicit about who they believe the customers are and their needs using tools such as personas, segmentation, or JTBD.
- This can then lead to conversations with stakeholders to align understanding.

Category: Collaborative

Even with the rise of AI tools that promise the ability for one person to design and deliver digital experiences, design is still usually a team activity. Designers need to collaborate not only with the teams delivering the design solution to customers but also with the business stakeholders who depend on the solution's success. Often for larger initiatives, there may be several designers who need to collaborate too.

There can be different understandings of what collaboration means and who is responsible for it. Although collaboration takes two or more parties to come together, the responsibility for ensuring it happens will often sit with the designer.

Gone are the days of designers accepting a brief, working in isolation for a few weeks or months, and delivering a final version for sign off. Designers are expected to collaborate throughout an initiative's design and delivery. There are clear benefits as issues and changes are identified early.

Important: Continuous collaboration also ensures that the design has a higher likelihood of acceptance, as stakeholders have bought into it over time.

Being collaborative is broken down into:

- Open
- Supportive
- Persuasive

Collaborative

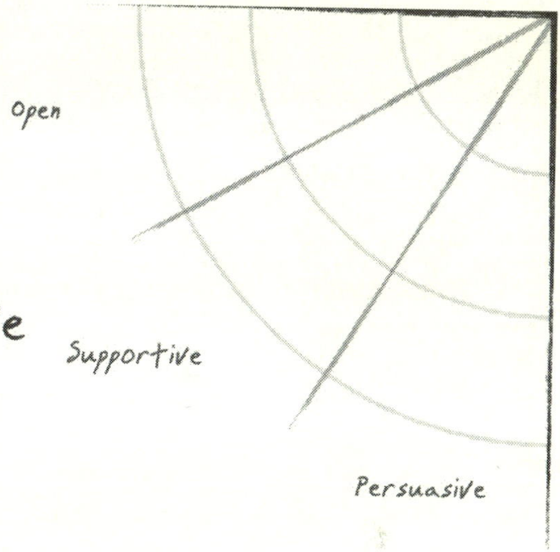

Collaborative: Persuasive

Being persuasive is an important part of collaboration. Persuasion requires an understanding of stakeholder views and needs and ensures that designs are focused on these. Often, immature designers will be frustrated that others do not accept or adopt their ideas. However, this is usually because they haven't been clear about why their idea is important and its impact.

Plan your communications

An important aspect of persuasion is a well-thought-through communication plan. As the old saying goes, 'You can't sell a secret'. The communication plan should identify who is important to persuade, the potential concerns, and the best-used channels. It is also important to consider that communication might be divided up over time rather than a one-shot chance to make a case. This will depend on the stakeholders and what resonates best with them.

Think deep, then speak

A persuasive designer is likely to have given their designs and ideas deep thought, been able to understand the essence of its strengths, and mitigate any issues. They have listened to early feedback.

Important: If they haven't adopted feedback, they will have clear reasoning that stakeholders can get on board with.

Right place, right time

The timing and nature of design sharing is critical for a designer to persuade others of the importance of adopting the ideas. It is much harder to persuade stakeholders of an idea if it doesn't resonate with them or is too late in the initiative.

Tell-tale signs of a mature designer:

- ☐ Proactively creates collaboration maps and integrates collaboration into their planning.
- ☐ Receives positive feedback from stakeholders.
- ☐ Customises reports and showcases to the audience.
- ☐ Proactively catches up with stakeholders to better understand their needs and concerns.
- ☐ Uses language that resonates with partners/stakeholders.

Potential barriers to scoring well

Lacks confidence.

Tactics to consider:

- Build up the designer's confidence using 'safe' forums to share ideas.
- Partner with the designer in initial collaboration sessions to provide a safety net.

Lacks presentation skills.

Tactics to consider:

- Create a development plan with the designer that includes presentation skills.
- Partner the designer with others who are skilled in this area.

Limited access to stakeholders.

Tactics to consider:

- Extend stakeholder maps to include alternates when stakeholders are unavailable.

Collaborative: Open

One benefit of a tertiary design education is that good schools would open students up to a variety of different areas. For example, a master's course in UX might cover human factors, learning theory, and psychological data collection methods, as well as UI design, service design, and so on. This is not to say that designers with tertiary education backgrounds are better, but they have often been exposed to more areas.

Important: Someone without advanced design education can (and often do) excel in design, especially if they are open to ongoing skills development and broadening their knowledge.

Asking for feedback isn't showing weakness

Openness to feedback is often strongest in early and late-career designers. There can be a period in between where there is a reluctance to gather feedback, which can limit growth. It's important for mid/senior designers to recognise and address this resistance to reach their full potential.

Carl Popper's falsification theory can be used as a basis for conversations to check openness. Asking a designer, 'What would it take to change your mind about your design?' reveals whether they're open to change or are fixed in their thinking.

Open to ideas from others

Being open means not only being open to feedback but also to ideas from others. Organisations often show tribal resistance to ideas from other teams. Some designers may view adopting external ideas as weakness or dismiss unfamiliar concepts too quickly. But reluctance to consider others' ideas often signals an overactive ego. Many top designers openly credit inspiration from teammates, other departments, or even outside organisations.

Look widely, think deeply

Being open to new ideas and ways of working enables the designer to adapt quickly and bring new approaches to the team. Reading widely, listening to podcasts, watching talks, and speaking with different teams are all signs of an open mind. These people are often seen sharing what they find and creating discussions around it to help them better understand these new areas.

Remember: It doesn't mean they need to jump on every new trend or constantly change their position, but they should be able to evaluate whether or not they need to change.

Tell-tale signs of a mature designer:

- ☐ Shares knowledge and ideas from beyond the world of design.
- ☐ Proactively seeks feedback on ideas and ways of working.
- ☐ Is able to change their mind when needed.
- ☐ Will embrace new ideas from other teams when needed.
- ☐ They are comfortable with ambiguity so will often develop several design solutions to gather feedback.

Potential barriers to scoring well

Overactive ego.

Tactics to consider:

- Focus on facts and solutions in discussions.
- Use frameworks like those in this book to avoid personal opinions when reviewing work.
- Set clear measurable goals that do not allow for deflection of fault if missed.

Time not allotted to gathering or acting on feedback.

Tactics to consider:

- Support the designer to plan early and ensure they hold regular collaboration and feedback sessions.
- Ensure the designer feeds back to stakeholders any barriers to implementing feedback, so they can help resolve them.

Lacks flexibility in business requirements.

Tactics to consider:

- Weigh up the impacts of changes to business requirements and escalate if required.

Collaborative: Supportive

Designers who are not supportive of each other signal that they are only focused on their own output and not that of the design practice.

Important: A customer or organisation has little care that there is one superstar in a design practice if the rest of the team is struggling to produce a cohesive, impactful user experience.

A designer who supports others is likely to have a deeper understanding of their work and will likely pick up new skills and perspectives from those they support. They are also likely to have a more rounded product understanding by being exposed to the thinking of other designers in the team.

Fragmented design structures should not be a barrier

When a design practice is fragmented (different designers are line managed in different parts of the organisation) or distributed (designers share line reporting but work embedded in different teams), collaboration can decline and rivalry may increase. However, mature designers look past shifting structures to value supporting colleagues wherever they sit, recognising the positive impact this brings.

Get your own stuff done first

Individuals not in leadership roles should balance supporting others without neglecting their own work, which is a common pitfall for some designers.

Tell-tale signs of a mature designer:

- ☐ Sets up regular catch-ups with other designers.
- ☐ Other designers often seek their advice or mentorship.
- ☐ In a fragmented or distributed design practice, they understand the work of other design teams well.
- ☐ Won't point fingers or diminish other designers and instead offers practical help.
- ☐ Will not undermine other designers when they aren't present.

Potential barriers to scoring well

Lacks capacity to support others.

Tactics to consider:

- Create practice-wide forums for support, such as critiques, that are worked into all designer's capacity planning.
- Work with the designer to ensure they're not overloaded.

Unaware of others' need for support.

Tactics to consider:

- Create forums for designers to share issues such as standups and shared planning boards.
- Encourage the designer to set up regular catch-ups with other designers.

Other designers do not want support.

Tactics to consider:

- Include mentoring in the designer's development plan.
- Have open and frank conversations about the importance of this for the practice culture.
- Include 360 feedback in any review process.

Category: Organised

Many who work with design teams note a lack of organisation compared to other practices. Often, team leaders manage design coordination while designers take a less active role.

Mature designers stay organised and ensure they have the time they need while not delaying others. Less mature designers may rush to deliver designs before fully understanding the scope, causing errors or delays.

Important: For design leaders, well-organised designers reduce task management and simplify reporting, freeing leaders to focus on higher-value work.

Being organised is broken down into:

- Proactive
- Efficient
- Timeline Focused

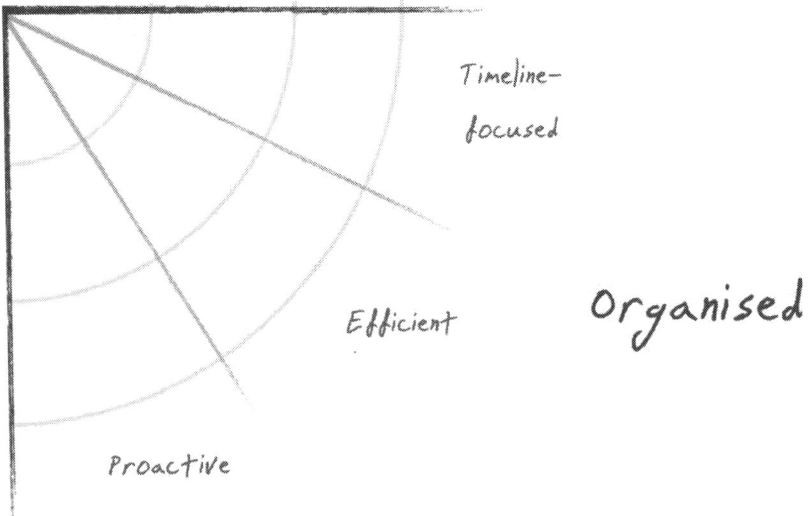

Organised: Proactive

A mature designer shouldn't need to be asked to plan their work; they should be able to break down their tasks and spot opportunities. Being proactive will also enable the designer to raise questions and highlight concerns early, while there's time to act.

Ensure you are considered in planning

Proactivity enables designers to contribute to initiative planning and estimation. Unlike development or operations teams, many designers feel overlooked during planning. By taking initiative, they can present their approach and timelines without waiting to be asked.

Known unknowns

Many unknowns arise early in design projects. Proactive planning helps designers address these with the team before they escalate and keeps design leaders informed of potential issues ahead of time.

Don't just plan early, do early

Being proactive means designers constantly seek to be better organised, such as creating stakeholder maps early, gathering past research, or adopting tools like design principles, systems, or project trackers to stay on top of their work.

Tell-tale signs of a mature designer:

- ☐ Will come to meetings well prepared.
- ☐ Will proactively gather information.
- ☐ Projects they work on typically include design timelines because they share their approach early.
- ☐ Stakeholders will always have a good understanding of when to expect design deliverables.
- ☐ Will volunteer rather than waiting to be asked.

Potential barriers to scoring well

Unaware of upcoming initiatives.
Tactics to consider:

- Work with the designer to find initiative roadmaps.
- Escalate with product/project management to ensure design is communicated promptly.

Left out of early planning/estimation sessions.
Tactics to consider:

- Assist the designer in forming strong relationships with project/product management.
- Educate teams about design processes and ways of working.

Reliant on project/product managers or leaders to set out timelines.
Tactics to consider:

- Assist the designer in breaking down and planning tasks to proactively present to project/product management.

Organised: Efficient

Rightly or wrongly, delivery teams often portray designers as being slow and potential road blockers. Effective designers regularly assess their processes to improve efficiency.

Looking for new and improved ways of working

Staying up to date with new tools and features is vital, as ongoing innovations help designers work more efficiently. While this may require some upfront time or cost, designers should recognise the efficiency benefits. In the evolving AI landscape, new tools regularly enable faster, higher-impact work.

Efficient designers also refine their methods continuously, using techniques like time boxing, templating, and pattern reuse.

Seeing opportunities

Seeing the bigger picture enables designers to quickly select the right tools or methods and engage others when needed.

Important: Efficiency for a designer isn't just about saving design time; it's about helping the whole team deliver quality into a customer's hands as quickly as possible.

Tell-tale signs of a mature designer:

- ☐ Frequently tries new tools/methods/refinements that save time.
- ☐ Proactive in retros and planning sessions, generating numerous ideas to improve efficiency.
- ☐ Discusses ways to save time for their colleagues.
- ☐ Doesn't waste time on methods that add little value.
- ☐ Can quantify time savings of different design approaches.

Potential barriers to scoring well

Unrealistic expectations of what is required.
Tactics to consider:

- Pair the designer with a more pragmatic designer as a mentor.
- Time box work, so the designer is forced to consider what is really necessary to maximise outcomes.

Works with teams that are not open to changing processes.
Tactics to consider:

- Share case studies from other teams the practice works with that have benefited from process changes.
- Work with the designer to be more persuasive in their interactions.
- Escalate with leadership.

Not thinking about the impact they have on others' workflows.
Tactics to consider:

- Include 360 feedback in any review frameworks.
- Ensure they attend retros with the teams they work with.

Organised: Timeline Focused

Designers sometimes skip basic planning methods such as breaking down steps, estimating, and ordering work. This results in missed tasks or last-minute rushes. It also makes it harder for design leaders to resource correctly.

Many methods and tools exist for planning design work. A key aspect is breaking down steps into a timeline, for example, when funding is needed, who must be consulted, and what deliverables are required and when.

Important: Many Agile planning methods identify the largest unknown or risky items and focus on these first, as it's better to tackle areas that might unpredictably impact timelines early.

Just because you are Agile doesn't mean you don't plan

Although some Agile purists avoid timelines, other teams need a clear order of events and estimated delivery dates. These may change with new information, but without a timeline, collaboration becomes difficult. Designers should also include contingency plans for extra revisions.

Be transparent

Mature designers openly share their plans, showing confidence and welcoming scrutiny that can build trust or uncover issues requiring further thought.

Tell-tale signs of a mature designer:

- ☐ Shares design planning early in projects.
- ☐ They're not often late delivering, and with few surprises.
- ☐ Receives positive stakeholder feedback.
- ☐ Demonstrates a calm demeanour throughout a project.
- ☐ Is able to quickly rework design timelines when new information comes to light.

Potential barriers to scoring well

Unable to break down tasks.
Tactics to consider:

- Support the designer in breaking down tasks to one or two days' worth of effort.
- Pair the designer with another who is skilled at task break downs.

High reliance on others with unclear timelines.
Tactics to consider:

- Instead of fixed dates, base effort estimates on others' delivery times.
- Escalate to get at least ballpark estimates from the other party.

Unwillingness to be held to a timeline.
Tactics to consider:

- Encourage the designer to put a contingency buffer on their estimates to gain comfort.
- Set a timeline for the designer if they are unwilling to.

Chapter 7: Worked examples of the individual designer conversation framework

To help build an understanding of how the conversation frameworks can be used by line managers and designers, we will run through three scenarios of different designers.

These include:

- Brief background about the situation
- How the designer scored against the framework
- Challenges they faced
- Areas they selected to tackle first
- The conversations they had using the framework

Scenario 1: Working on key gaps to progress a designer's career

Background

- Has been with the organisation for two years.
- In a mid-level role.

Problem to solve

The designer wanted to move into a senior role but needed to address gaps in design quality. Despite strong stakeholder relationships, their work often required basic reworks, highlighting the need for targeted development to support career progression.

Current success measures

- Positive feedback from stakeholders
- Positive feedback from delivery teams
- Design quality is mixed. Will often require substantial reworking

Forum for discussions

- Monthly professional development catch-ups

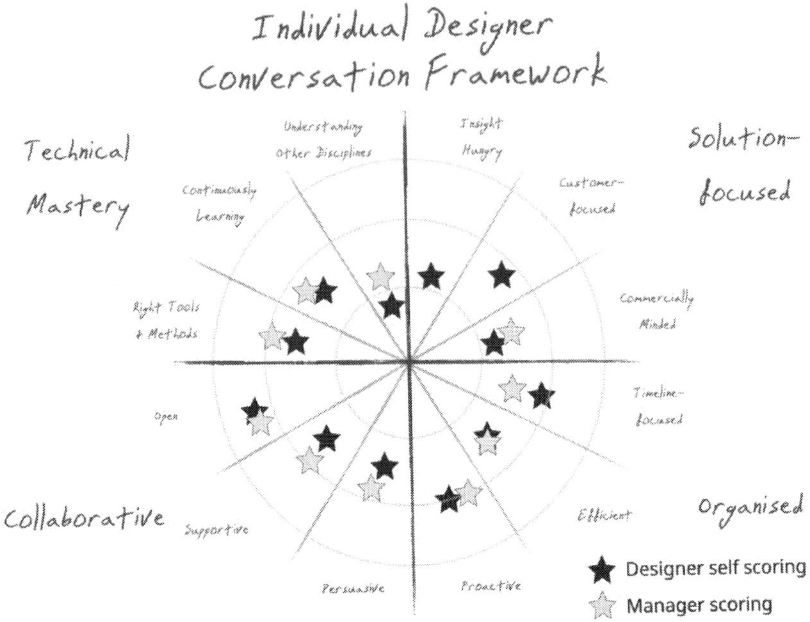

Approach taken

Introduction of the designer conversation framework

- The manager introduced the designer conversation framework to the team, explaining that it would be used to guide future professional development conversations.
- The designer completed their section of the framework and shared it with the manager, who also gathered supporting evidence to contribute.
- As a result, the designer expressed interest in progressing to a more senior role.

Alignment and focus areas

- The designer and their manager were generally aligned in their scoring.
- During their monthly professional development catch-up, they discussed the scores and explored reasons for any differences.
- They also talked about the designer's career goals, including their aspiration to move into a more senior role.
- Together, they identified three key areas for the designer to focus on and planned a set of development activities.
- These included connecting with colleagues to share knowledge, working on projects that provided exposure to relevant skills, and participating in external training.

Ongoing reflection

- In each career development meeting, they referred back to the framework to reflect on progress and update it as needed.

Focus area for next quarter

- Solution Focused: Insight Hungry
- Technical Mastery: Understanding Other Disciplines
- Organised: Timeline Focused

Scenario 2: Limited self-awareness regarding areas for growth

Background

- Has been with the organisation for five years.
- In a senior design role for the past six months.

Problem to solve

Since the designer's promotion to a senior role six months ago, mixed feedback from stakeholders and delivery teams has raised concerns about some areas of under-performance.

Current success measures

- Mixed feedback from stakeholders.
- Mixed feedback from delivery teams.
- Design quality is satisfactory.

Forum for discussions

- Fortnightly 1-to-1s.

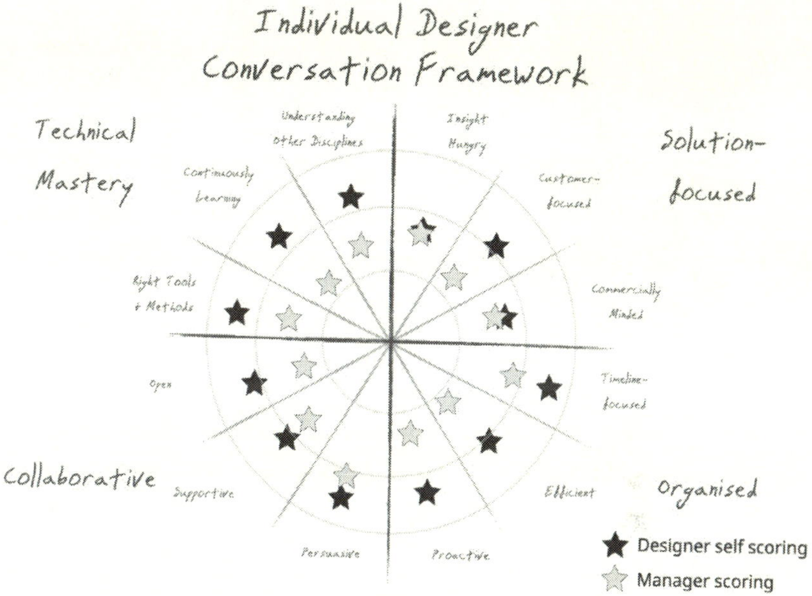

Approach taken

Introduction of the designer conversation framework

- The manager introduced the conversation framework during a team day and asked each team member to complete it ahead of their next 1-to-1 meeting.
- The designer hadn't completed the framework in time for their scheduled catch-up, so the manager postponed the meeting to allow them time to complete and submit it.
- Once received, the manager added their own input, including evidence, and brought the updated framework to their rescheduled 1-to-1.

Feedback and performance observations

- The manager had received mixed feedback from colleagues who had worked with the designer.
- They had also observed a decline in some areas of performance since the designer's promotion six months earlier.

Differences in assessment scores

- The designer was surprised to learn that there were significant differences between the designer's and manager's scores in key areas.
- The manager explained the lower scores were due to a lack of supporting evidence and stakeholder feedback.
- In some cases, the designer was able to provide additional examples in key areas, allowing the manager to revise their assessment.
- For the lower-scoring areas, they explored what might be holding the designer back. Some of the challenges required the manager's support to improve ways of working with a couple of partner teams.

Agreement and ongoing review

- They agreed on three focus areas for the designer to take responsibility for and clearly defined what success would look like in each.
- In every subsequent 1-to-1, they used the framework to reflect on progress and assess whether the designer's development efforts were shifting the dial on the agreed outcomes.

Focus area for next quarter:

- Collaborative: Open
- Solution Focused: Customer Focus
- Organised: Efficient

Scenario 3: New early-career designer

Background

- New employee.
- In a junior role.

Problem to solve

As a new junior employee, the designer needed structured support to get started, but there wasn't yet enough evidence for a formal assessment. The challenge was to guide early development in a meaningful way while setting the foundation for future growth.

Current success measures

Not established at this early stage.

Forum for discussions

- Two weeks of regular catch ups as they got up to speed.
- Fortnightly 1-to-1 catch ups.

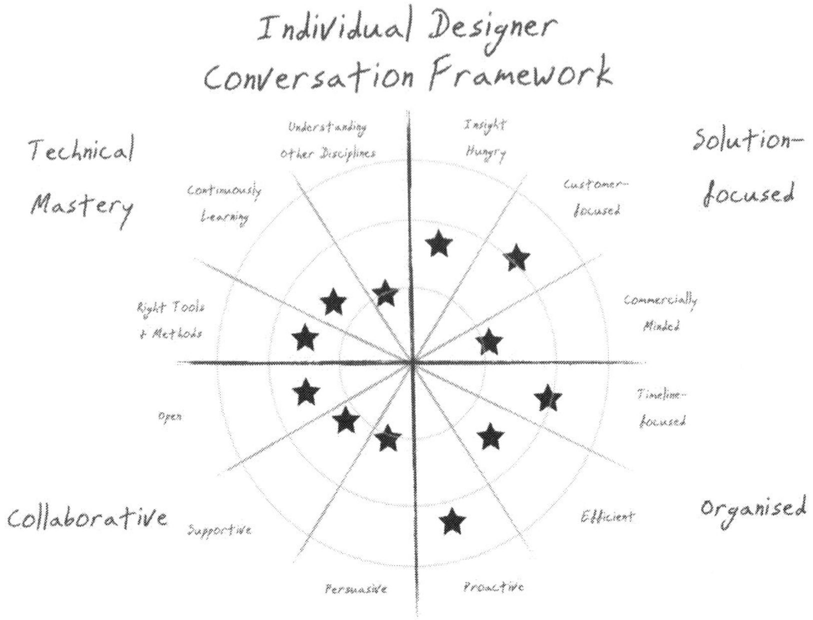

Approach taken

Introduction of the design conversation framework

- The team had been using the designer conversation framework for a while.
- The framework was introduced as part of their team induction when a new junior designer joined.
- They were asked to complete it within the first couple of days, so their manager could use it to help build a career development plan.

Focus on self-assessment

- As the designer was new, the manager wasn't in a position to provide scoring, so the focus was on self-assessment.

Early development conversations

- In their early 1-to-1s, the manager used the framework to guide conversations about areas the designer could begin focusing on once they were up to speed.
- For this initial period, they agreed on two areas to focus on.
- Much of the development plan centred around creating a mentoring relationship with a more established designer on the team.
- The new designer spoke with the established designer, explaining their desire to improve in an area where the established designer was strong.
- The established designer took them under their wing, sharing their approach and allowing the new designer to shadow them for the first few weeks.

Revisiting the framework

- After a few months, the manager revisited the individual designer conversation framework with the junior designer.
- With enough evidence gathered from their work, the manager was now able to complete their own assessment.
- This led to a more meaningful discussion about any differences between the self-assessment and manager scoring, helping to clarify expectations and guide the next stage of development.

Focus area for next quarter:

- Technical Mastery: Right Tools & Methods
- Solution Focused: Commercially Minded

Chapter 8: Applying the design conversation framework

The framework should be used consistently throughout the design cycle, even when a design is shared for the first time. This helps set clear expectations for others reviewing the design and provides insight into any trade-offs that have been made.

As the design progresses toward completion, the framework offers a way to assess the design's overall health. For example, later in the project, it's expected that feasibility and research would score better compared to the early stages

Important: When engaging with stakeholders, the framework can guide discussions about trade-offs to ensure everyone is comfortable with decisions made.

Since many design trade-offs are often overlooked, the framework helps bring them to light. It also encourages stakeholders to look beyond the design's aesthetic appeal and understand the risks being carried into the delivery phase.

The framework has four categories:

- User-friendliness
- Short & Long-Term View
- Alignment
- Proof Points

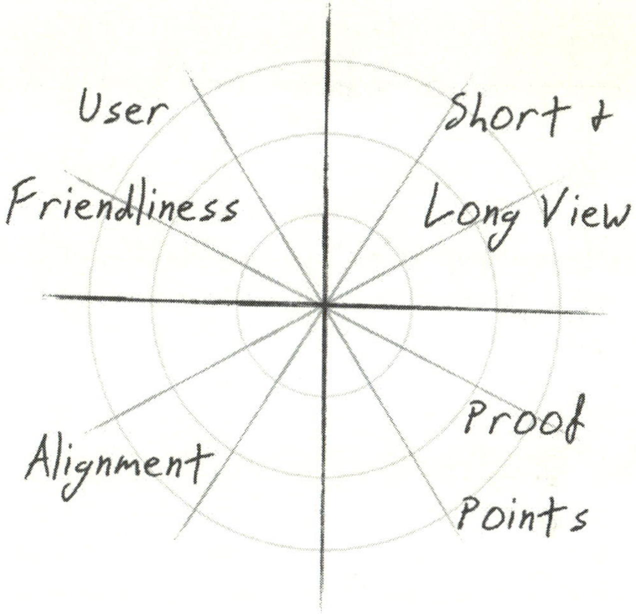
User
Friendliness
Short &
Long View
Alignment
Proof
Points

Category: User-friendliness

In the early days of digital interaction, the term 'user-friendly' was widely used. However, it has largely fallen out of many designers' vocabulary: not because they don't believe it's important for their designs to be welcoming and frustration-free for users but because the term feels overly simplified. Despite this, 'user-friendly' remains a well-understood and universal term that's easy to communicate to stakeholders and customers. It helps keep the focus on user connection with the system.

Ultimately, the end user judges design quality. If they don't find the design enjoyable or easy to use, it misses the mark. Enjoyment doesn't necessarily mean a 'woo-hoo' moment. Often, it simply means that users can complete their tasks with ease.

Some designers focus solely on user-friendliness. While it's extremely important, it's not the only consideration.

Important: If a design scores poorly in user-friendliness, strong performance in other areas won't compensate because users simply won't like it.

Poor scores in this category often stem from inexperienced or absent designers, significant technical constraints, or ineffective management.

User-friendliness is broken down into:

- Appealing
- Easy to Use
- Useful

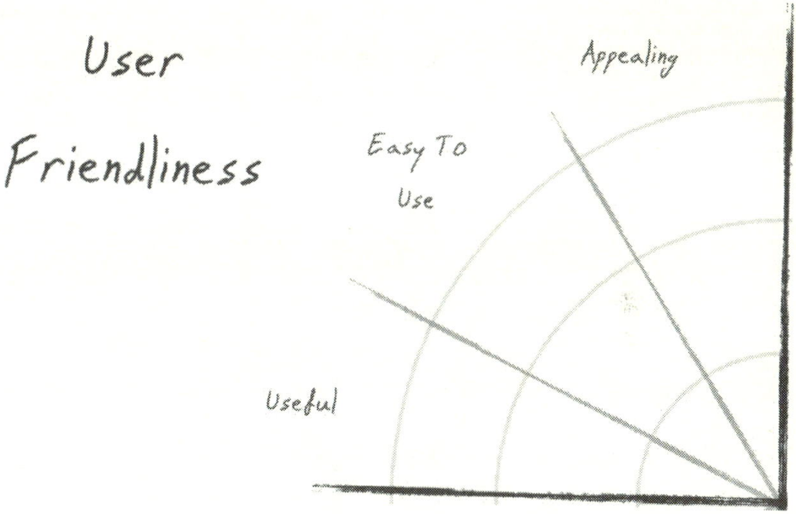

User-friendliness: Appealing

Visual appeal has been shown in studies to increase the perception of ease of use and enjoyment. This then reflects well on the organisation's brand and sets users up well for interacting with other parts of the services offered. Some very successful organisations have built a reputation for being easy to use with visually appealing interfaces despite scoring poorly on key usability metrics. Let's not name names, but I am sure you can all think of one or two.

Note: Visual appeal is a chosen focus due to the current dominance of visual interfaces. This should be swapped out if the design focuses on other senses, such as audio.

It's mostly objective

The topic of visual appeal is often clouded by subjectivity and personal preference. Even CEOs who should know better sometimes insist on using their favourite colours or fonts.

Important: Much of visual appeal is grounded in objective principles, drawing on psychological research and validated through customer testing.

Situation of use impacts visual appeal

Visual appeal is situational, for example, something that looks great on desktop doesn't necessarily translate well to mobile. Yet, a surprising number of designers and managers still prioritise desktop design at the expense of mobile usability.

Context also matters

Where and how a design is interacted with can greatly influence its effectiveness. Is it part of a broader ecosystem of interfaces? Is it being used in specific environmental conditions like dark mode on an operating system or outdoors in bright sunlight? Designs that account for these situational factors demonstrate a deeper level of consideration and are more likely to stand the test of time.

Brand focused

Fonts, colours, white space, animations, hierarchy, interactions, and more should all be well thought out and culturally appropriate to the intended audience. These elements may be bespoke to the organisation or reused from third-party frameworks. Importantly, they should align with the brand ethos, which has varied rules and standards at different organisations. Brand guidelines can extend well beyond visual elements to include emotional tone and broader experiential considerations.

Unfortunately, alignment isn't always straight forward. Some brand guidelines are developed with a focus on advertising without considering how they translate to product or interface design and may even introduce accessibility issues.

Important: Neither the design nor the customer should be penalised for a lack of forethought in branding decisions.

Tell-tale signs of a mature design:

- ☐ Stakeholders do not have suggestions for aesthetic changes.
- ☐ Brand and accessibility guidelines are adhered to.
- ☐ User testing demonstrates customers' approval of the aesthetics.
- ☐ The design does not degrade on popular devices.
- ☐ The design will not quickly feel dated.

Potential barriers to scoring well

Brand guidelines do not translate well.
Tactics to consider:

- Address issues with the brand team.
- Diverge from brand guidelines but only when absolutely necessary.

Technical limitations.
Tactics to consider:

- Meet with technical experts to understand available options.
- If the proposed changes are high value, collaborate with the technical team to run a spike and assess potential limitations.
- Determine whether a similar outcome can be achieved within those technical constraints.

Top-down design direction.
Tactics to consider:

- Conduct user testing to ensure the proposed direction doesn't lead to negative outcomes.
- Leverage a design system to maintain consistency and reduce the need for custom aesthetic changes.

User-friendliness: Easy to Use

Imagine you're on a trip with five friends. Four are outspoken and know exactly what they want to do. One is quieter and doesn't share much about their preferences. It's likely the quieter friend will get less attention and have less say in what the group does, not because their needs don't matter but because they're less visible.

Ease of use often ends up in the same position on digital projects. It competes with louder, more defined demands like business goals, security requirements, legacy concerns, and technical limitations. As a result, usability is often treated as flexible or something to adjust or sacrifice when compromises are needed. Despite its importance, it's frequently the first to be deprioritised in the noise of complex delivery projects.

Test for ease of use

Testing with end users and gathering feedback can happen at any stage of design and development/release. Testing early and often helps fill knowledge gaps and uncover blind spots, giving the team confidence that the design is truly usable.

But it isn't one-size-fits-all. A design that works well for experienced users might be confusing for new ones. For example, returning online customers may have a good understanding of domain-specific language and the current layout, while first-time users might struggle.

Important: Designers need to understand the size and needs of each group they're serving.

While it's important to make the experience accessible to less experienced users, this shouldn't come at the cost of frustrating larger or higher-value customer segments.

It should be easy for all to use

Accessibility is a key aspect of ease of use, but it's often misunderstood or narrowly applied. Many teams focus on technical compliance, such as screen reader support for users with significant impairments. While this is essential, it's only the starting point.

The value of an accessible design goes beyond compliance; it improves user experience's overall quality, often resulting in cleaner code, better structure, and more thoughtful design decisions. It also serves a much broader group than many realise.

Accessibility can be especially challenging for design teams because some of the most critical factors, like technical implementation and content, may sit outside their control. Designers still play a key role in supporting accessibility activities by partnering with the teams who influence these areas.
Does the design have guidelines about what standards must be followed to pass testing, or are there copywriting guidelines so that the experience doesn't degrade as the content evolves?

Important: Accessibility depends on a shared understanding of users and their contexts. Without that insight, accessibility risks becoming a compliance checkbox rather than a driver of inclusive, effective experiences.

Tell-tale signs of a mature design:

- ☐ Passes accessibility guidelines and accessibility user testing.
- ☐ Positive feedback from target customers.
- ☐ Cognitive load to complete popular tasks is low.
- ☐ Low likelihood of customer support required for most users.
- ☐ It is easy to use on all popular devices/situations.

Potential barriers to scoring well

Lacks buy-in from others regarding the importance of ease of use.
Tactics to consider:

- Use case studies to underline the impact.
- Gather customer feedback about ease of use impacting the choice of product vs competitors.
- Ensure ease of use is highlighted by senior leaders.

Differing views on the target customers and their needs.
Tactics to consider:

- Use accepted tools such as JTBD, personas, or segments to agree on customer groups and needs.
- Share insights from customers about their needs.

Technical or budget limitations.
Tactics to consider:

- Reference commercial and ROI impacts of limiting ease of use.
- Work with delivery teams on short spikes to truly understand what is possible.

- Look for alternative design solutions that have similar ease of use but are more feasible.

Impacts on other areas of the product.

Tactics to consider:

- Leverage code and design reuse tools, such as design systems, to reduce the cost of making design updates across the product.
- Create a long-term plan for phased rollouts of design improvements across all areas.

User-friendliness: Useful

Have you ever put your belongings into storage? It's a difficult task. You have to decide what's worth keeping and what can go, factoring in future use, original cost, emotional value, or simply how much space you have. Then, when you unpack it later, how often do you ask yourself, 'Why did I   this?'

Design initiatives are often like this. Teams hold onto legacy features and rules for all kinds of reasons. Over time, these decisions pile up, making the product harder to navigate and less useful for users. Like an overstuffed storage unit, the clutter gets in the way of what actually matters.

Less is more

The first question any designer should ask is, 'What customer need is this solving?'. If the answer isn't clear, why spend time and resources progressing the work?

Important: This question should be applied to the designer's own ideas, not just those proposed by others.

The most successful products are built around useful features. However, that usefulness must be defined from the user's perspective, not based on a hunch from a designer or stakeholder. In the age of AI where there are fewer barriers to releasing new products and features, asking questions about the usefulness of ideas early is more important than ever.

Understanding the customers and their underlying needs is critical to designing products fit for use. What matters most is consistently grounding design decisions in real user insight to boost work quality and prevent low-value ideas from making it through.

Over time, products tend to accumulate more and more features. The logic is often under the belief that more means better. But every interface has a tipping point, where the experience degrades by adding more, no matter how useful the feature. Each new and old feature must earn its place.

Important: When adding a new feature, it's worth asking, 'Is there something less valuable that can be removed?'

Not every necessary feature is useful

Ideally, customer needs and business goals are aligned, but that's not always the case. Some features may be necessary for commercial or regulatory reasons. These should be clearly identified during design reviews to ensure they're included for the right reasons, not due to outdated or incorrect assumptions.

Refactoring isn't just for code

Designers should be mindful of how many features a given design pattern can support before it starts to break down. Once that limit is reached, the design may need to be restructured, much like developers refactor code to maintain performance and scalability.

Tell-tale signs of a mature design:

- ☐ Data gathered from user interest testing (for example, false door tests or Kano surveys).
- ☐ Positive customer feedback in user testing.
- ☐ The feature set is aligned with the product/design/company strategy.
- ☐ Commercial goals are linked to user goals.
- ☐ Customer journeys and feature sets are compelling, i.e., they pass the 'pub test'.

Potential barriers to scoring well

Feature bloat adds inherited complexity to new features.
Tactics to consider:

- Work with stakeholders to review the value of all features and set a retirement plan for underperforming ones.
- Refactor designs to give a more appropriate feature hierarchy.

Misguided top-down direction.
Tactics to consider:

- Work with senior leaders to expose the insights and thinking used by the team.
- Run rapid trials to gather data on the pros and cons of different approaches.

Feature requirements are set before customer needs are understood.
Tactics to consider:

- Work with leaders to refactor ways of working so that requirements are not set in stone before critical information is gathered.

- Escalate with leadership and discuss strategies to lead with rapid customer insights where possible.

Strong emotional connection to features.
Tactics to consider:

- Create an agreed framework for assessing the value of ideas, no matter who creates them.
- Run research to review the impact of all ideas.

Features are solely focused on short-term commercial goals.
Tactics to consider:

- Work with business leaders to ensure initiatives have balanced long and short-term goals where appropriate.
- Build business cases for features with longer-term payoffs.

Shallow research that focuses purely on customer wants rather than needs.
Tactics to consider:

- Create assessment frameworks for research and only use research that meets minimum standards.
- Work closely with researchers/analysts to ensure needs are understood.

Category: Alignment

Like leaving handprints in wet cement, many designers want to make their mark. It's understandable; there's pride in ownership, and it helps when showcasing personal impact, especially for those seeking career progression.

Remember: Customers don't care who designed what. And business stakeholders care more about efficiency and customer feedback than designer ego.

Alignment isn't always easy, especially when teams work in silos or don't share plans openly. But it's still possible in key areas. Where it isn't, that misalignment should be a deliberate, informed decision made with clear understanding.

Designing for alignment takes more skill than just going your own way and ignoring the work of others. Alignment with other teams requires more effort upfront, thoughtful planning, and a willingness to compromise. It pays off by minimising friction, reducing rework, and preventing the need to escalate issues up the chain.

True alignment comes from collaboration and openness. Informed teams can offer better input and are more likely to adapt when needed, making for a smoother, more efficient design process.

Alignment is broken down into:

- Feasible
- Impacts Goals
- Shared Patterns

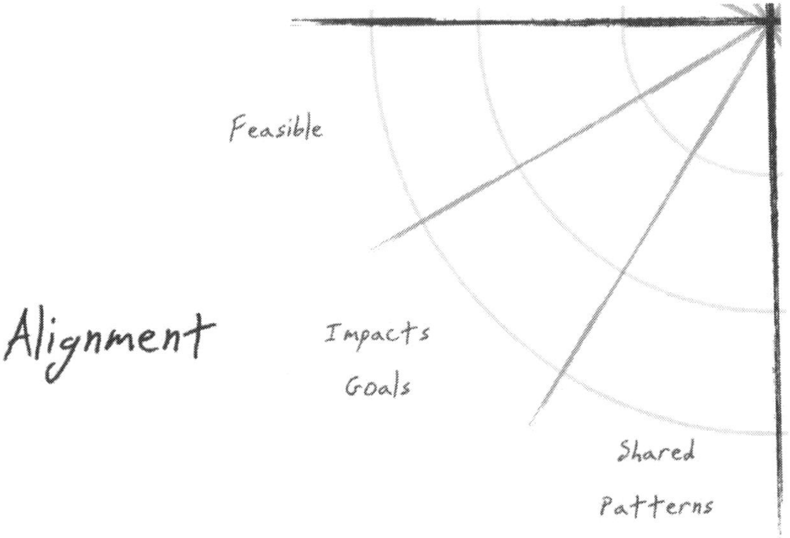

Alignment: Feasible

Why is feasible important for alignment?

A design isn't real until it reaches the customer. Until then, it's a plan with a set of intentions, passed on to delivery teams to bring to life. If the design isn't feasible for those teams, it's unlikely to be delivered. This is a common frustration for both designers and business leaders who signed off on a design that never ships, resulting in wasted effort and commercial investment.

Important: Feasibility goes beyond technical development; it also depends on the teams who will maintain the experience over time. These teams have their own budgets, systems, and processes to maintain the design's feasibility. This doesn't mean designs can't challenge or even reshape those constraints, but it does mean those impacts need to be identified early and factored into the overall delivery's timing and budget.

Designs should consider feasibility early

Feasibility is often considered too late in the design process. While it's important to give teams creative freedom early on, delaying feasibility checks can lead to overinvestment in ideas that are unlikely to meet ROI commitments.

Feasibility and assumptions are bad bedfellows. The more a design is built on untested assumptions, the lower the confidence in its feasibility. That's why designers need to clearly identify those assumptions and prove/disprove them as early as possible. Transparent, early testing is key to ensuring feasibility and avoiding costly course corrections later.

Tell-tale signs of a mature design:

- ☐ High levels of confidence from the whole project team.
- ☐ Project timelines are clear.
- ☐ During review sessions, there are few design revisions required.
- ☐ Any assumptions used to create the design are well documented.
- ☐ The project team has all the information they need to estimate the effort required to implement the design.

Potential barriers to scoring well

Many unknowns and assumptions.

Tactics to consider:

- List out the assumptions and work with stakeholders to prove or disprove them.
- Alternatively, add contingency to design timelines.

Lacks access to those who can speak to feasibility.

Tactics to consider:

- Find an alternative representative who can speak to feasibility or add contingency and escalate.

Alignment: Impacts Goals

For international soccer, 2002 was a notable year. Not only did Brazil win its fifth FIFA World Cup title, but in Madagascar, a very different soccer record was set. During a game featuring the team AS Adema, one player scored 146 own goals. AS Adema lost the match 149-0, setting a world record and making the game infamous worldwide.

A well-thought-out design must aim for the right goal and clearly articulate its impact on organisational objectives. For example, how does the design increase revenue, reduce cost to serve, or increase satisfaction? If the design doesn't impact those organisational goals, then is it worth the investment?

Important: At this stage, designers must grasp the commercial objectives tied to the approved initiative. Achieving this often involves working closely with product management and other business stakeholders.

Ongoing costs should not be ignored

Reducing the cost to serve customers is a common business goal that designers frequently overlook. Designing for efficiency and simplifying the customer experience should be priorities, since most customers value straightforward services. However, cutting costs too aggressively can degrade quality. As with all commercial goals, there is often a balance to ensure the customer, employees, and organisation benefit.

How will goals be measured?

A design's true impact often isn't clear until after launch, and even then, measuring it can be difficult. Tools like A/B testing offer valuable insights, while trials and prototypes can help build reliable hypotheses about impact before significant resources are committed.

What if the goals can't be reached?

If the design solution isn't impacting the right goals or isn't delivering sufficient impact, it likely needs to be reconsidered and its return on investment may not justify continuing. This could be due to an underestimated scope or because the design isn't targeting the appropriate customer segments or key metrics.

Tell-tale signs of a mature design:

- ☐ Stakeholders provide positive feedback.
- ☐ The design has measurable goals.
- ☐ Trials and/or A/B testing are used to understand the impact on the initiative goals.
- ☐ Design features are linked to initiative goals during run-throughs.
- ☐ Assumptions about how the design will impact goals are clearly documented.

Potential barriers to scoring well

The initiative doesn't have clearly stated goals.
Tactics to consider:

- Escalate with leadership to understand what they had in mind when prioritising the initiative.

The initiative's goals have changed.
Tactics to consider:

- Refactor the design and provide a new timeline.

Business requirements do not match the initiative goals.
Tactics to consider:

- Estimate the impact on goals and escalate with leadership to gain clarity.

Alignment: Shared Patterns

Shared design patterns have always played a role in design, whether in physical products with reusable components or in print where typesetters used standardised layouts. The benefits of shared patterns, such as interoperability, familiarity, and speed, often outweigh the benefits of using fully custom solutions.

The level of shared pattern reuse varies, but a well-considered design that doesn't use a shared pattern must answer: 'Is the additional effort worthwhile?'. If so, 'Can others reuse this new pattern?'.

Important: If a pattern can't be reused, there must be enough unique value to justify the extra effort to create and maintain it and for the user to learn how to use it.

Assist delivery teams

Shared patterns help delivery teams stay aligned. For example, using shared policies and code bases makes it easier for people to move between teams and products. They provide a consistent frame of reference and allow more people to contribute ideas and suggest improvements.

Shared patterns aren't just about visuals

Sharing patterns carry risk, especially when changes made by one team to a shared pattern can unintentionally impact others. That's why many organisations establish clear procedures and governance rules to manage them responsibly.

Design principles are also a form of shared patterns.

Important: A design team that aligns on core principles are more likely to produce cohesive, consistent work. For example, a team may agree on principles that focus on accessibility, balance, or user support.

When those principles are embedded in the work, any design, regardless of who creates or updates it, will align with the team's broader direction.

Tell-tale signs of a mature design:

- ☐ The design is cohesive with the rest of the product and aligns with existing design principles.
- ☐ Has clear business reasoning when not using a shared pattern.
- ☐ Incorporates reusable patterns that can be adopted by other teams or projects.
- ☐ Implementation teams can quickly implement the design with minimal explanation.
- ☐ Current customers are able to use the new product/service with minimal effort.

Potential barriers to scoring well

Legacy design and code issues make reuse difficult.
Tactics to consider:

- Investigate the opportunity for a design system.
- Work with stakeholders to plan for code and design refactoring in the near future.

Stakeholders have strong preferences to be different from other areas of the product.
Tactics to consider:

- Present current and future costs/timing impacts of any changes to the stakeholder group.

The need to do something unique is clear.
Tactics to consider:

- Design with future reuse in mind.

Category: Proof Points

When buying a house, many buyers commission a house survey, even after seeing the property and hearing assurances from the seller and real estate agent that, 'It's fantastic, take my word for it!'. When a designer is selling their design to stakeholders, often with costs comparable to buying a house, should stakeholders simply take their word for it?

Remember: A strong design must show its credentials. Providing relevant proof points helps stakeholders understand the rationale behind decisions. Stakeholders are far more willing to take risks when the design is backed by solid proof.

Important: The amount of proof required depends on the risk involved.

High-cost projects or those that could negatively impact many customers demand a greater burden of proof. Conversely, quick changes with minimal risk require less evidence.

Proof points are broken down into:

- Reviewed
- Researched
- Competitor Comparison

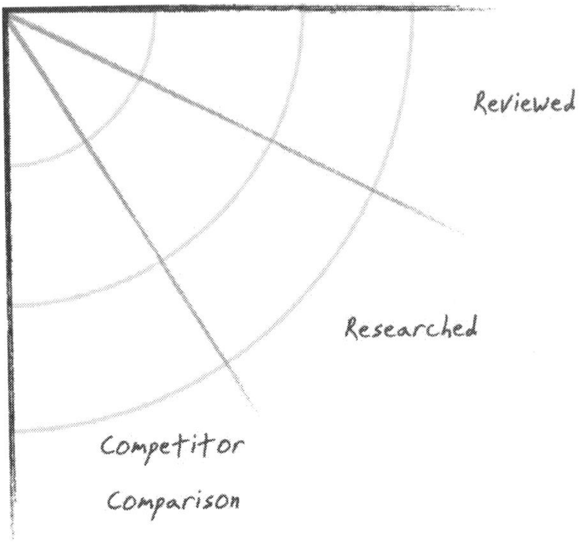

Proof Points

Proof Points: Reviewed

Diverse thinking is a key ingredient for successful design. Design teams often hold critiques specifically to provide designers with constructive feedback from their peers. However, it's equally important to seek input beyond the immediate team. Bringing a broad range of perspectives to design reviews can reveal new insights and spark fresh ideas.

Important: Without review from others, designs risk overlooked issues that may result in costly rework down the line.

During reviews it is important separate personal feelings from feedback. Biases exist on both sides between reviewers and recipients. Recognising these biases helps designers discern which feedback to act on and determine the appropriate response.

Take Ownership & Follow Up (TOFU) with feedback

One commonly overlooked part of the review process is the follow up. Designers may take feedback on board and explore it further, but the outcomes can vary. It might not lead to a change, it could result in a different adjustment than what was suggested, or it may be directly implemented. Whatever the outcome, it's important to close the loop by updating the person who gave the feedback.

Important: This shows their input was understood and appreciated, even if it wasn't adopted, and helps build trust, making future feedback easier to collect.

Tell-tale signs of a mature design:

- ☐ Final designs are of no surprise to the team and stakeholders.
- ☐ Design project plans include stakeholder/feedback maps and factor in feedback sessions and design iterations.
- ☐ Feedback sessions are documented along with any resulting actions.
- ☐ Feedback is reviewed and assessed to understand its validity.
- ☐ Stakeholders have confidence that their feedback has been heard and understood.

Potential barriers to scoring well

Lacks time to action customer/stakeholder feedback.

Tactics to consider:

- Factor feedback and design iterations into project planning.
- Create a longer-term view of the design that incorporates the feedback but in a later iteration.

Team/stakeholder feedback is not constructive.

Tactics to consider:

- Ensure the team/stakeholders have the information they need to understand any constraints.
- Conduct heavily structured feedback sessions/critiques to ensure targeted feedback.
- Seek to understand the intent behind the feedback, as it may reveal gaps or overlooked issues in the design.

Proof Points: Researched

Thirty years ago, giving every employee access to a computer was considered a luxury. Fifteen years ago, it became a justifiable investment. Today, it's a basic expectation, whether the employee is stacking shelves, managing accounts, or driving a truck.

Design research has followed a similar path. Once seen as slow, expensive, and specialised, it's now faster, more affordable, and widely accessible. So, why hasn't it earned the universal acceptance in business as essential to good design? In many organisations, the time and cost to gather user insights are now low enough that they don't meaningfully affect timelines, and using them often accelerate reviews and approvals. Yet, outdated perceptions of research linger.

This isn't to say all research is fast and cheap. Like top-tier computers, some research is costly and worthwhile in complex or high-stakes situations.

Important: In most cases today, design research's speed, accessibility, and affordability mean the greater risk lies in skipping it altogether.

Research trade-offs

Research always involves trade-offs between quality, time, and budget. The right level of research will vary by initiative. Factors like the need for corroborating evidence, the scale of the change, and the associated risk all help determine how much research is required.

Important: Any research used to support a design must meet a minimum standard of reliability. Both designers and stakeholders should clearly understand the level of rigour behind the insights informing a design.

Choose wisely

A wide variety of insights and data sources can support a design. But with so many options available, it's critical to match the approach to the initiative's risk level, investment, and timeline. In many cases, insights are gathered post-launch through trials, A/B testing, or standard release monitoring. However, if post-delivery feedback is a key part of the strategy, plans must be in place to act on that feedback.

Remember: Collecting data serves little purpose if there's no path for addressing the issues it reveals.

Be aware of bias

Designers and reviewers must be mindful of the many biases that can influence research. Confirmation bias is especially common in user research. Expected results are often accepted without much scrutiny, while unexpected outcomes tend to face more questioning. To strengthen reliability, research should draw on multiple data points and take a scientific approach, testing to disprove a design hypothesis, not simply to confirm it.

Tell-tale signs of a mature design:

- ☐ Research/insights are readily available to support the design.
- ☐ Team and stakeholder confidence in the design research is high.
- ☐ Pre- and post-release feedback gathering is planned and accounted for in planning for the design's future iterations.
- ☐ Several research and insight methods are used to minimise bias.
- ☐ All research used has been reviewed to ensure rigour.

Potential barriers to scoring well

Not enough time to gather insights and/or data.
Tactics to consider:

- Leverage appropriate historical insights.
- Run rapid user testing/surveys.
- Highlight additional timeline risks of unforeseen issues.

Lack of access to insights and/or data.
Tactics to consider:

- Investigate the availability of alternative data sources.
- Budget for establishing your own access to insights/data.
- Escalate to leadership to gain access.

Research is carried out but ignored.
Tactics to consider:

- Continue to reference ignored insights throughout the project.

Proof Points: Competitor Comparison

At the 1968 Summer Olympics in Mexico City, Dick Fosbury shocked the world by winning gold in the high jump using a revolutionary technique. Fosbury learned his technique by studying others. While he didn't invent the back flip, he knew why it was rarely used. He recognised that new soft crash pads allowed for a new style of jump, the 'Fosbury Flop', which all successful high jumpers soon adopted.

Fosbury's success came from deeply understanding his competitors' techniques and spotting an opportunity they hadn't exploited. His rivals quickly followed once they saw the advantage. Fosbury succeeded by thoroughly understanding what was happening in his field. Successful future high jumpers succeeded by quickly adopting his winning technique.

In design, there are often competitors to learn from, though not always. Strong designs take lessons from what has or hasn't worked for others. Problems arise when teams simply copy competitors without critical thought. Many designers are told to 'just do what X did' without understanding why that approach was taken or whether it's appropriate for their situation.

Avoid survivor's bias

It's crucial to learn not only from competitors' successes but also from their failures and to share these insights openly with stakeholders. Survivor's bias often leads organisations to mimic only the winning strategies of others, overlooking where those strategies may have fallen short. This narrow focus risks copying tactics that don't guarantee success, as unsuccessful projects may have used the same approaches.

Important: To truly understand what drives success, both wins and failures must be examined.

Incorporating competitor solutions into research helps designers uncover what customers appreciate, what frustrates them, and what factors could sway them to choose your product over others.

Use tools to reduce costs

Competitor analysis has traditionally been a time-intensive process, requiring detailed tracking of features and user journeys. Frequent updates by competitors made it challenging to stay current with design changes. Today, AI agents can automate many of these tasks, allowing teams to dedicate more time to thoughtful, strategic evaluation of competitors.

Tell-tale signs of a mature design:

- ☐ It has been compared to a range of competitors.
- ☐ Customers prefer the design to that of competitors.
- ☐ A compelling marketing plan can be developed to effectively compete with competitors.
- ☐ Competitor customers can effortlessly transition to this product or service.
- ☐ Stakeholders are well-informed about competitor similarities and differences and the reasoning behind them.

Potential barriers to scoring well

Stakeholders demand that the design copy competitors.

Tactics to consider:

- Conduct customer research to determine whether competitor features should be copied.
- If a competitor has a good solution, don't avoid adopting it just for ego's sake.

There are no competitors to compare to.

Tactics to consider:

- Look for similar industries to review or solutions in other territories.

The overhead of constantly reviewing competitors is too large a burden on the team.

Tactics to consider:

- Use AI tools to automate the scraping of competitors.

Category: Long and Short View

Today, workplaces are filled with people who need to know the big picture to be able to plan for the future. Business leaders deal with this every day as they balance their investments. Many business leaders will think in the McKinsey 'three horizons' model. So, when they see a new design, they won't just be thinking about the impact of the first release.

Senior stakeholders will often ask a design team not only to design for the next iteration but also what the experience might be in the future. This is sometimes a challenge when delivery teams are being directed to only look to the next iteration.

The short-term view of the design solution should be fairly accurate, but the long-term view can be looser. This is to accommodate for unknowns that will show up between now and the future. The long-term view should be flexible to allow for new insights surfacing and changes in market conditions.

Set expectations

Being clear on the expectations of the first and later iterations is extremely important, as poor expectation management can lead to inaccurate views of project failure. Providing future views is fraught with opportunities for partner teams to become annoyed with the design team, so aligning with project and product management is important here to ensure stakeholders and delivery teams receive a consistent message.

Long & short view is broken down into:

- Within Budget
- Measures in Place
- Future View

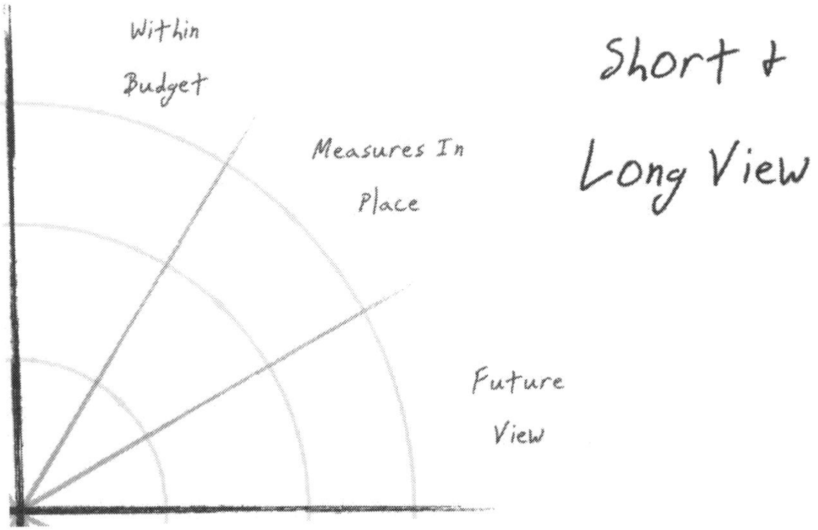

Long & Short View: Within Budget

Understanding the budget available for an initiative is crucial. A budget may include development time, licensing fees, and ongoing maintenance costs. Stakeholders weigh the cost against the potential return when prioritising projects. If a proposed design exceeds the budget, it may be prudent to prioritise a lower-cost alternative.

Overly simplified designs are common, often driven by pressure to deliver quickly and demonstrate progress. But if simplification goes too far, commercial goals are at risk of being missed. In theory the Minimum Viable Product (MVP) from the PO provides this guidance, but in practice it is a term often misused and thus ignored in some organisations.

Important: Regardless of the terminology, the key principle remains: A minimum level of value must be achieved relative to the effort invested.

Flexibility in budgets

Budgets can often be adjusted if the design delivers greater value. When evaluating ROI, higher returns generally justify increased investment (i.e. budget). This flexibility is usually reserved for future iterations, so presenting stakeholders with a clear plan for both the initial release and subsequent updates helps clarify the full investment needed.

Tell-tale signs of a mature design:

- ☐ Every feature has an associated return or commercial goal.
- ☐ The cost of development, maintenance, etc., is discussed when the design is showcased.
- ☐ There are few, if any, design assumptions that could lead to additional work later on.
- ☐ The design's budget is revisited only if additional value is uncovered.
- ☐ Business stakeholders express a high level of confidence that the design will be delivered within budget.

Potential barriers to scoring well

The initiative doesn't have clear commercial goals.

Tactics to consider:

- Use research to understand the initiative's customer value.
- Escalate to leadership to gain clear goals.

The initiative was prioritised without clear investment information, so it is unknown if the design is within budget.

Tactics to consider:

- Raise with leadership to understand their perspective on the initiative's scope and value.

Stakeholders push for low-value, high-cost features.

Tactics to consider:

- Work with product/project management to map the Desirability, Viability, and Feasibility (DVF) of all features, providing stakeholders with a clear overview of feature prioritisation.

Long & Short View: Measures in Place

Establishing clear success measures or key indicators ensures both the immediate effectiveness and long-term potential of a design are understood. These measures should be identified early in the design phase, alongside plans for how they will be gathered. Success measures and design go hand-in-hand. Without a way to evaluate success or failure, progressing the design lacks purpose.

Some metrics, like click-through or conversion rates, require analytics tags or automated reporting. Others, such as customer or employee satisfaction, rely on softer measures like post-release surveys. While these surveys happen after launch, responsible teams should begin planning them early to ensure timely execution.

It's also important to consider strategic, binary success measures for designs that may not deliver standalone results but set the stage for future success. For example, a flexible menu structure might not yield immediate gains but supports long-term product evolution. Success here is measured by alignment with broader design goals.

Important: Long-term success metrics differ from standard engagement measures by focusing not only on current usage but also on potential growth. An initial release might serve one customer segment, with later iterations expanding to new groups.

Together, short- and long-term success metrics provide valuable insights that allow designers to iterate and enhance future versions.

Tell-tale signs of a mature design:

- ☐ Each success measure has an agreed target to reach.
- ☐ Measures are either in place or there is a clear plan to implement them by go-live.
- ☐ Stakeholders see the value in the success measure chosen.
- ☐ Stakeholders understand when and how they will receive reports on the design's impact.
- ☐ It's agreed that the initiative will not be marked as 'complete' until success measures and feedback are reported on.

Potential barriers to scoring well

Lacks infrastructure to measure success.

Tactics to consider:

- Build a business case for investment in success measures.
- Escalate with leadership to highlight the uncertainty of success due to lack of measures.

Reporting on success measures is not prioritised.

Tactics to consider:

- Escalate with leadership to gain prioritisation.
- Look for easier solutions to measure success.

Success measures are not clear.

Tactics to consider:

- Workshop success measures with the project team.
- Engage with leadership to clarify the rationale behind the initiative's prioritisation so that success measures can be appropriately aligned.

Long & Short View: Future View

Successful designers often begin with a clear long-term vision and then work backward to shape the next iteration, helping avoid costly rework and future problems. Design patterns often have optimal use cases, for example, some menu types work well for shallow menu structures common in early product iterations but struggle with larger structures that may come later. The designer should discuss the pros and cons of choosing something optimal for the first iteration but suboptimal for future versions.

However, there can be conflicting pressures. Often, senior stakeholders want to understand long-term possibilities while delivery managers worry that higher expectations might affect scope. Design artifacts should clearly differentiate what is for the current iteration and what is planned for the future.

Maintain flexibility

The long-term vision should remain adaptable, as new insights will shape the evolving experience. Future design concepts should include disclaimers noting this uncertainty, and designers must be ready to adjust their plans based on emerging information.

Creating multiple future scenarios is common practice. This allows designers to explore how the design might shift depending on factors like targeting a broad mass market versus a specialised customer segment.

Tell-tale signs of a mature design:

- ☐ There are multiple future design options depending on how the product evolves.
- ☐ A future-facing view of the design has been established, including information on its evolution's potential value.
- ☐ Stakeholders and delivery managers are comfortable with the future view.
- ☐ The next iteration released to customers is understood by all involved.
- ☐ The future view is adaptable.

Potential barriers to scoring well

Not enough time to look to the future.

Tactics to consider:

- Time box future vision work to fit within the available project timeline.

Heavy pushback from product/delivery managers about looking at the future view.

Tactics to consider:

- Take extra effort to set expectations about the status of the future view.

Future view is seen as being owned by other teams.

Tactics to consider:

- Clarify how far into the future stakeholders are comfortable exploring and design accordingly.
- Escalate with leadership to highlight the design team's capability to envision the future.

Chapter 9: Worked examples of the design conversation framework

To help build an understanding of how designers can use conversation frameworks, we will run through three scenarios of different designs.

These include:

- Brief background about the situation.
- How the design scored against the framework.
- Challenges they faced.
- Areas they selected to tackle first.
- The conversations they had using the framework.

Scenario 1: Design is 80% complete with major reworks since the concept stage

Background

- The designer had been working on the initiative over several iterations.
- Early research indicated the initial features were not compelling to customers.
- This led to an expanded scope aimed at better meeting user needs.

Problem to solve

The original design lacked features that resonated with users, putting the initiative's goals at risk. The challenge was to align the design with customer needs within a limited budget while demonstrating progress toward increasing visits and satisfaction.

Current success measures

- Increasing visits (goal of 5%+ increase).
- Customer satisfaction (goal of 3%+ increase)

Forum for discussions

- Design critique.
- Stakeholder showcases.
- Product owner meetings.

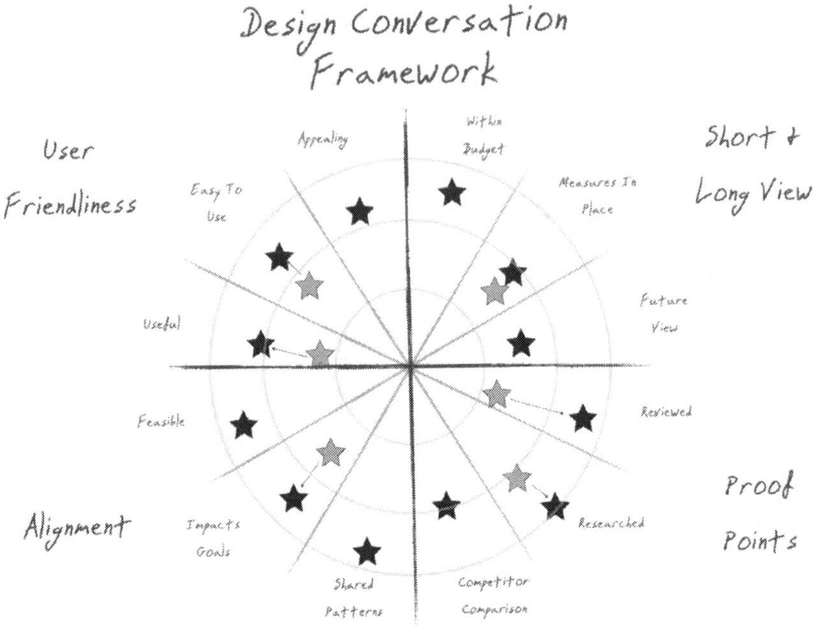

Approach taken

Initial concerns and scope limitations

- The designer had conducted co-design sessions with stakeholders and developers and came away concerned that the current scope was too limited to deliver features customers would find useful.
- As a result, the design risked failing to meet key initiative goals of increasing customer visits and satisfaction.

Introduction of the design conversation framework

- To explore this further, the designer used the design conversation framework to map the current design's strengths and weaknesses.

- They shared this with the product owner (PO) in a regular meeting.
- The PO shared similar concerns but noted the budget, specifically developer hours, was limited.
- They also discussed the lack of a long-term view (6-12 months) of the design's evolution.
- Given the many unknowns at this stage of the initiative, they agreed to postpone future-state design work until after customer feedback is received post-launch.

Team collaboration and research support

- Later that day, the designer brought the challenge to their team's weekly design critique.
- Using the design conversation framework, they mapped the design's current state and received additional ideas and an offer of support for further research.

Identifying improvements and securing budget

- Through this follow-up research and another co-design workshop with developers and stakeholders, the team identified low-cost improvements that would better meet user needs and contribute to the success metrics.
- Since developers had been involved in the co-design process, it was easy to gather feasibility input, which helped the PO secure a modest budget increase.

Demonstrated progress

- In the second round of design critique, the team responded positively. The updated design, again mapped using the design conversation framework, clearly demonstrated the progress made.

Focus areas

- Short & Long View: Future View (once feedback from live product is received)
- Proof Points: Compared (if competitors copy the ideas)

Scenario 2: Design that leverages competitor success 30% complete

Background

- A competitor launched new features and was promoting them heavily.
- Stakeholders worried this was impacting their Share of Voice (SOV).
- A designer was tasked with creating mock-ups of similar features for an upcoming meeting.
- No research.
- Patterns reused.
- The delivery team has not been consulted at this stage.

Problem to solve

At the first design review, the design lacked research and delivery input, creating uncertainty about user value and technical feasibility. The challenge was to validate assumptions and risks before proceeding.

Current success measures

- Share of voice (no target provided).
- Sales (2%+ increase).
- Customer satisfaction (80%+ with new features).

Forum for discussions

- Stakeholder meetings.

- Feasibility meetings.

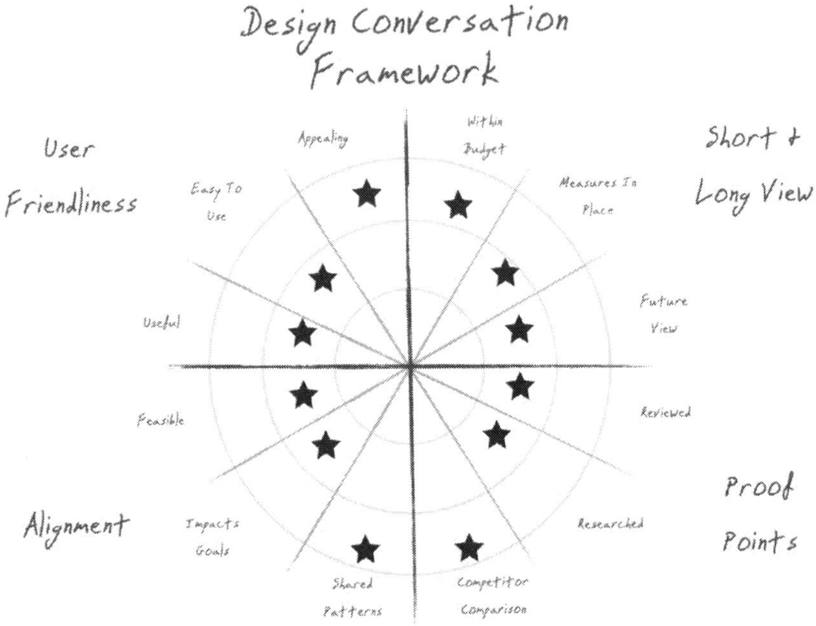

Approach taken

Using the design conversation framework

- To support the conversation, the designer used the design conversation framework to highlight both the strengths of the design and the key unknowns.
- This was shared with stakeholders alongside visual mock-ups.

Stakeholder feedback and concerns

- Stakeholders recognised that while the features were aesthetically appealing, there were several unknowns and risks.
- They agreed that before prioritising development, more information was needed on feasibility and customer value.
- That said, if the features proved feasible within budget, implementation was likely mainly to respond to marketing pressures.

Outcome and impact

- Stakeholders were impressed by the quality of the designs and the depth of thinking demonstrated through the design conversation framework.
- After launch, the features had a negligible impact on sales, and customer satisfaction was low, which aligned with the designer's expectations.

Focus areas

- Alignment: Feasible
- Proof Points: Researched
- User Friendliness: Useful

Scenario 3: Low risk but low impact design idea 50% complete

Background

- Customer research identified a gap for users who prefer not to contact support when encountering problems.
- In response, a designer created a mock-up and worked with an innovation team to develop a prototype for testing.
- This well-researched, low-risk initiative is the first step in a multi-iteration solution and lays the foundation for future design patterns.

Problem to solve

The team needed to validate that the prototype effectively reduces support calls and meets user needs, ensuring assumptions are tested early to guide future design iterations.

Current success measures

- Sales conversion (15%+ increase).
- Lifetime customer value (10%+ increase).
- Customer satisfaction (75%+).

Forum for discussions

- Innovation team meetings.
- Design critiques.
- Stakeholder showcases.

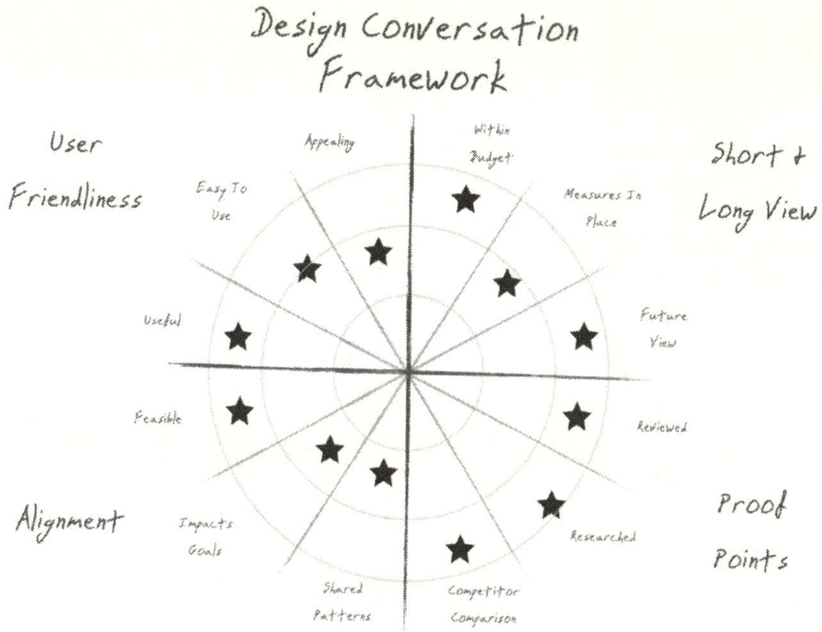

Approach taken

Using the design conversation framework

- The designer brought the mock-ups to the design team's critique session.
- They used the design conversation framework to highlight a lack of shared design patterns since the current design system didn't fully support the solution.
- The team suggested reusable patterns and noted that some of the new patterns might be beneficial for their own future work.

Stakeholder presentation and alignment

- When the design was presented to stakeholders, they were impressed by the clear alignment with customer needs identified in the research.
- While the new features weren't expected to move the current business metrics the team was tracking, they did address a key business objective: reducing call centre contacts.
- As a result, the design was green-lit for prototyping, and the team agreed to revisit their OKRs to better support this type of initiative in future.
- This included beginning to track customer support volume to measure the design's impact.

Ongoing refinement and broader impact

- While feedback is gathered on the prototype, the designer plans to refine the visual design and develop reusable patterns that can benefit the broader design team

Focus areas:

- Short & Long View: Measures in Place
- Alignment: Impacting Goals
- User Friendliness: Appealing

Chapter 10: I've completed a framework, now what?

Design and design practices don't succeed in a vacuum; they thrive through critical thinking, alignment, and quality communication.

Whether you're working at the design practice, designer, or design level, using your completed framework is about turning insights into action.

Design practice, designer, and design conversations next steps

- Practice-level: As a team, identify one or two low-scoring areas. Collaboratively build a strategy to improve them. Including your team early improves buy-in and outcomes.
- Designer-level: Introduce the framework before using it in professional development conversations. These discussions can be sensitive, so avoid surprises.
- Design-level: Bring the framework into a retro or design critique. Let the designer fill it in beforehand to frame the design review.

Measure what matters

The frameworks help you identify where you are and where to focus next, but they don't measure impact. To understand and demonstrate the effect of your actions, set clear success metrics before you begin.

Remember: Three good base measures for a design practice, designer, or design are output speed, quality, and confidence. But you may have your own that resonates with your organisation.

These metrics ensure your conversations lead to meaningful action and build trust when referenced again after improvements have been made.

Start small, share thoughtfully

You've completed a framework. Now think about:

- How will you share this with your team and stakeholders?
- Where can it fit into your existing processes?
- Who needs to be involved in the conversation?

Design conversations work best when they evolve. Start small and build iteratively, just like with good design.

Find the right forum

No one wants an extra meeting if it can be avoided. Consider whether your design conversation can fit into an existing meeting or if it's better shared as a document or short video.

Choose the format that best suits your audience and intent.

Afterwards, check in to confirm:

- Was the message clear?
- Are the next steps understood?

The goal isn't just to share, it's to ensure alignment.

Understand your audience

As with design, effective conversations start with knowing your audience.

- What style of conversation works best for them?
- Are they information focused or story focused?
- What existing knowledge and preconceptions do they have?
- What biases are they likely to have?

Important: When in doubt, tailor your message to the most influential people in the room.

Always clarify what action you want from them: input, support, investment, or change.

Use the visuals

Assuming visual accessibility isn't a barrier, show the framework but don't talk about everything in it. The visuals help people see the broader picture and understand how different areas connect rather than needing to 'waffle on' about everything in the framework.

Important: The framework supports your conversation, but it's not the conversation itself. The real focus should be on your specific situation and the key areas you want to address.

Be authentic

Trust matters. Leverage what you're known for:

- Are you a strong presenter? Work the room.
- Do you prefer 1-to-1 chats? Suggest a catch-up.
- Are you a stats nerd? Bring the data.
- Are you good with humour? Use it to make your point.

Being authentic allows you to draw on the trust you've built over time. But if you're starting fresh and don't have an existing relationship or reputation, take a moment to establish one. Briefly share why this conversation matters to you and what you're hoping to achieve. That one minute of honest context helps set the tone and makes the rest of the conversation flow more naturally.

Be consistent

The frameworks in this book give you a shared foundation for design conversations. Use them consistently across teams and stakeholders to keep your message clear and aligned.

Important: Consistency builds trust. If your focus constantly shifts, people may disengage. A steady, repetitive approach helps others stay on board and makes it easier for them to carry your message into conversations you're not part of.

When ideas clash

Design conversations may challenge people's beliefs, ideas, or egos. That's normal. Don't frame it as your idea versus theirs.

Remember: Use the framework to neutrally explore multiple options.

If change is needed, frame it as an evolution, not a rejection. If your practice improvements require changes to the way others currently work, use the discussion framework to highlight why change is required and how current practices need to be built upon.

You're building on what exists, not tearing it down. There may be deep-rooted beliefs in what you are looking to change, or your change may result in additional work by others that needs to be considered.

Important: Influence, don't demand.

Stay focused

Design conversations can easily drift, whether from over-explaining or competing agendas. Stay focused on your intended topic and park unrelated issues for follow up later.

If the timing is wrong and other priorities dominate, it's better to pause and revisit later than push through. Choosing the right forum and understanding your audience helps avoid this, but knowing when to pivot protects your credibility.

Don't just talk. Act!

This book has focused on the essential conversations between designers and leaders to build alignment. But talk is cheap. Actions speak louder.

Now that you've earned trust through well-designed conversations, it's time to deliver results. Focus where it matters, and seize opportunities to improve outcomes.

After taking action, use the frameworks to tell the story of your achievements.

Conversations create alignment, but action builds credibility.

Chapter 10 Summary
Start small
Understand your audience
Measure impact
Choose the right forum
Follow up with action!

Chapter 11: About me

My motivation for this book

Like many others, I am dyslexic. This affects how I communicate both in writing and verbally, which has made me think deeply about communication. Early on, I learned that no matter how skilled or knowledgeable you are, poor communication can undo your work.

This inspired me to create clear frameworks to help communicate design and design leadership more effectively. After talking with design and business leaders, I discovered that many face communication challenges. Design leaders struggle with their communication approach, while business leaders often find it hard to understand how design teams operate.

My background

Growing up in the UK in the '80/'90s meant witnessing one of the biggest shifts in design and technology. From the introduction of the home computer and the early internet to the rise of Agile, smartphones, service design, and AI, the evolution of customer experience has been constant.

Computing was a natural choice for me at university, with strong job prospects and growing relevance. The course, shaped by the British Computing Society, combined software engineering with business, law, project management, and human-centred design, which really resonated with me. I went on to complete a Master's in Human-Centred Design.

My career spans both large traditional organisations and fast-moving digital companies, giving me a broad perspective. After starting my career in Europe, I moved to Australia in 2010, I built the design and research practice at a mid-sized sports betting company. With new ownership and big ambitions, we quickly led the market through superior products and user experience.

Next, I worked at a large traditional retailer transitioning to ecommerce. Here, I faced different challenges: low digital maturity, large commercial goals, and the need to start a design practice from scratch. With a lean, customer-centric approach, the team I built helped make our digital platforms a major driver of company profits and omni-channel customer satisfaction.

In both roles, success came from connecting design to business value. Projects had to show commercial impact, not just customer insight. That meant balancing usability with ROI and speaking the language of business, not just design.

Not every project went to plan. For example, during an Agile transformation, a review of design effectiveness exposed a gap. We had research, resources, people, and maturity frameworks in place, but the message didn't land. The approach lacked clarity and alignment. It was a valuable lesson. Influence needs simplicity, measurable outcomes, and a clear link to strategy. With that focus, our team achieved significant commercial and customer success.

Outside work, I founded and ran the Design Leadership meetup group in Melbourne for several years. This gave me valuable insight into the shared and shifting challenges design and business leaders face across organisations. Since then I have been having regular catch ups with design leaders and executives to understand their stories of success and failure.

These experiences shaped the thinking behind this book: That clear, thoughtful conversations are at the heart of great design and great leadership.

Drop me a line

I am keen to hear how you get on with designing design conversations. Any thoughts and feedback are much appreciated. You can find me on LinkedIn, Medium, and Substack.

Made in the USA
Monee, IL
05 November 2026

67387538R00136